Piotr Skrzypacz

Finite element analysis for flows in chemical reactors

Piotr Skrzypacz

Finite element analysis for flows in chemical reactors

Südwestdeutscher Verlag für Hochschulschriften

Impressum / Imprint
Bibliografische Information der Deutschen Nationalbibliothek: Die Deutsche Nationalbibliothek verzeichnet diese Publikation in der Deutschen Nationalbibliografie; detaillierte bibliografische Daten sind im Internet über http://dnb.d-nb.de abrufbar.

Bibliographic information published by the Deutsche Nationalbibliothek: The Deutsche Nationalbibliothek lists this publication in the Deutsche Nationalbibliografie; detailed bibliographic data are available in the Internet at http://dnb.d-nb.de.

Coverbild / Cover image: www.ingimage.com

Verlag / Publisher:
Südwestdeutscher Verlag für Hochschulschriften
ist ein Imprint der / is a trademark of
OmniScriptum GmbH & Co. KG
Heinrich-Böcking-Str. 6-8, 66121 Saarbrücken, Deutschland / Germany
Email: info@svh-verlag.de

Herstellung: siehe letzte Seite /
Printed at: see last page
ISBN: 978-3-8381-3805-3

Zugl. / Approved by: Magdeburg, Otto-von-Guericke-Universität, 2010

Danksagung (Acknowledegment)

Der Spiritus Movens dieser Arbeit ist mein langjähriger Betreuer und Vorgesetzter Prof. Dr. Lutz Tobiska. Für seine kritischen Anmerkungen, Geduld, aufregenden Diskussionen, schöpferische wissenschaftliche Zusammenarbeit auf dem Gebiet der Finiten Elemente Methode und der Numerik der singulär gestörten Randwertprobleme und für die Unterstützung meiner Assistentenstelle gebührt ihm ein besonderer Dank.

Mein spezieller Dank gebührt der Deutschen Forschungsgemeinschaft (DFG), welche mein Forschungsprojekt (FOR 447) im Rahmen der interdisziplinären Forschergruppe der "Membranunterstützten Reaktionsführung" finanzierte. Dem Sprecher dieser Forschergruppe Prof. Dr.-Ing. Andreas Seidel-Morgernstern bin ich zu Dank für die ausgezeichnete wissenschaftliche Zusammenarbeit verpflichtet.

Die Umsetzung der vorgestellten Methoden erfolgte dank der vorgeleisteten Arbeit der MooNMD-Entwickler. Hervorheben möchte ich an dieser Stelle die praktischen Programmierfähigkeiten von Prof. Dr. Gunar Matthies und ihm für die hervorragende wissenschaftliche Zusammenarbeit danken.

Sehr dankbar bin ich ebenfalls apl. Prof. Dr. Friedhelm Schieweck für seine konstruktive Kritik, bewegende Diskussionen - speziell seine Hinweise zur diskreten inf-sup Bedingung sowie die Eröffnung neuer Zukunftsperspektiven für mich.

Bei Dr. Walfred Grambow bedanke ich mich für seine Betreuung der Hard- und Software an meinem Arbeitsplatz. Priv.-Doz. Dr. Bernd Rummler verdanke ich, dass er durch seine Liebe zur Funktionalanalysis und zu klassischen Methoden der mathematischen Physik meine Interessen an diesen Fächern geweckt hat.

Es soll auch die Zusammenarbeit mit Priv.-Doz. Dr. Matthias Kunik nicht vergessen bleiben. Er hat mir neue Blickwinkel auf die Methoden der klassischen Analysis eröffnet und mich durch seine Erfahrungen bereichert.

Frau Sybille Enzmann aus der naturwissenschaftlichen Abteilung der Magdeburger Universitätsbibliothek schulde ich Dank für die reibungslose Bereitstellung der gesuchten Fachliteratur.

Allen ArbeitskollegInnen und Arbeitskollegen an der Otto-von-Guericke-Universität und meinen Professoren aus der Studienzeit an dieser Universität, die hier nicht einzeln erwähnt werden können, danke ich für die konstruktive Zusammenarbeit.

An meine Freundin Dipl.-Inf. Sandra Kutz, die mich beim Widerfahren alltäglicher Probleme begleitet, richtet sich mein persönlicher Dank. Ihre Lebenserfahrung und ihre Liebe zum Sport und zu Naturwissenschaften gab mir neue Motivation zur wissenschaftlichen Arbeit.

Zum Schluss möchte ich mich bei meinen Eltern für ihre umfangreiche Unterstützung und unermüdliches Bemühen bedanken.

Dipl.-Math. Piotr Skrzypacz
Institut für Analysis und Numerik
Otto-von-Guericke-Universität Magdeburg
Universitätsplatz 2, 39106 Magdeburg

Contents

1 Introduction

The question of a sustainable use of fossil fuels becomes nowadays more and more important due to the shrinking natural resources. The modern reactor chemistry takes advantage of using catalysts in order to increase yields, selectivity of products or to achieve the better control over reaction processes. In the last decade computational fluid dynamics turned out to be a convenient method for development of more efficient chemical reactors and for prediction of their behaviour. We present in this work a mathematical model for flows in packed bed membrane reactors and a simple model equation of convection-diffusion-reaction type.

The aim of this work is to elaborate robust numerical schemes which can be applied to subproblems resulting from models of chemical reactors, like flow or transport equations of diffusion-convection-reaction type. Moreover, we provide rigorous numerical analysis of our new schemes and test them by solving academic problems as well as by simulating flow behaviour in the packed bed reactor. A good source for readers more interested in questions of reactor modelling and in new trends in the membrane reactor engineering is a practical approach book [71].

The underlying work is devoted to mathematicians interested in scientific computing and finite element analysis of problems related to mass or heat transfer in chemical reactors as well as to engineers dealing with simulations of chemical reactors. The novel methods presented in this work are addressed also for graduates being engaged in scientific computing.

All of the chapters which deal with the numerical analysis are complemented by computational results. Our new schemes have been successfully implemented into the object oriented in-house finite element package MooNMD, see [46], and the computations have been performed on an ordinary Linux workstation.

Notation Throughout the work we use the following notations for function spaces. For $m \in \mathbb{N}_0$ and bounded subdomain $G \subset \Omega$ let $H^m(G)$ be usual Sobolev space equipped with norm $\|\cdot\|_{m,G}$ and seminorm $|\cdot|_{m,G}$. We denote by $D(G)$ the space of $C^\infty(G)$ functions with compact support contained in G. Furthermore, $H_0^m(G)$ stands for the closure of $D(G)$ with respect to the norm $\|\cdot\|_{m,G}$. The counterparts spaces consisting of vector valued functions will be denoted by bold faced symbols like $\boldsymbol{H}^m(G) := [H^m(G)]^n$ or $\boldsymbol{D}(G) := [D(G)]^n$. The L^2 inner product over $G \subset \Omega$ and $\partial G \subset \partial\Omega$ will be denoted by $(\cdot,\cdot)_G$ and $\langle\cdot,\cdot\rangle_{\partial G}$, respectively. In the case $G = \Omega$ the domain index will be omitted. Throughout the paper we denote by C the generic constant which is usually independent of the mesh and model parameters, otherwise dependences will be indicated. ◇

The following two examples are intended as small appetisers illustrating the problem of ensuring stability and accuracy of numerical solutions.

1.1 Why to stabilise Galerkin scheme?

Let us consider the two-point boundary value problem for singularly perturbed one-dimensional convection-diffusion equation which has been discussed in [68]

$$\left.\begin{aligned} -\varepsilon u'' + bu' &= 0 \quad \text{in } \Omega = (0,1), \\ u(0) &= 0\,, \\ u(1) &= 1\,, \end{aligned}\right\} \tag{1.1}$$

whereby the perturbation parameter is $\varepsilon > 0$ and the convection field $b > 0$ is constant. The exact solution of (1.1)

$$u(x) = \frac{e^{bx/\varepsilon} - 1}{e^{b/\varepsilon} - 1}$$

exhibits a boundary layer of width $O(\varepsilon/b)$ near to $x = 1$ if $0 < \varepsilon/b \ll 1$. Galerkin discretisation by piecewise-linear finite elements over a uniform grid

$$x_j = jh\,, \quad j = 0, \ldots, M\,, \quad M \in \mathbb{N}\,, \quad h := \frac{1}{M}\,,$$

leads to the following tridiagonal linear system for the unknown nodal values of the discrete solution u_h

$$\left(-\frac{\varepsilon}{h} - \frac{b}{2}\right) u_{i-1} + \frac{2\varepsilon}{h} u_i + \left(-\frac{\varepsilon}{h} + \frac{b}{2}\right) u_{i+1} = 0\,, \qquad i = 1, \ldots, M-1\,,$$

whereby $u_i := u_h(x_i)$, $u_0 = 0$, $u_M = 1$. Assuming that $2\varepsilon \neq bh$, the solution vector is given by

$$u_i = \frac{\left(\dfrac{1+Pe}{1-Pe}\right)^i - 1}{\left(\dfrac{1+Pe}{1-Pe}\right)^M - 1}\,, \quad i = 1, \ldots, M-1\,, \tag{1.2}$$

where

$$Pe := \frac{bh}{2\varepsilon}$$

is the Péclet number. $Pe > 1$ implies $\frac{1+Pe}{1-Pe} < 0$ and consequently the discrete solution u_h exhibits unphysical oscillations. They can be eliminated if the mesh size h gets small so that $Pe < 1$. If $\varepsilon/b \ll 1$ the mesh refinement leads to big systems which are, especially in higher dimensions, infeasible from the numerical point of view.

1.2 Taking advantage of superconvergence phenomena

The following small example illustrates the superconvergence phenomenon that has been discussed in [72]. Let us consider the two-point boundary value problem

$$\left.\begin{aligned} -u'' &= f \quad \text{in } \Omega = (0,1), \\ u(0) &= 0\,, \\ u(1) &= 0\,, \end{aligned}\right\} \tag{1.3}$$

whereby f is assumed to be sufficiently smooth. We use the same equidistant grid as in the example (1.1)

$$\mathcal{T}_h = \{K = (x_i, x_{i+1}),\quad i = 0,\ldots,M-1\}\,,\quad h_i = x_{i+1}-x_i = h = \frac{1}{M},\quad \forall\, i = 0,\ldots,M-1\,,$$

and look for the discrete solution u_h in the space of continuous piecewise linear polynomials over $\mathcal{T}_h$ and with zero boundary conditions

$$V_h = \{v_h \in C(\bar{\Omega}):\quad v_h|_K \in P_1(K),\quad v_h(0) = v_h(1) = 0 \quad \forall K \in \mathcal{T}_h\}$$

The finite element discretisation of problem (1.4) reads as follows

Find $u_h \in V_h$, such that

$$(u_h', v_h') \;=\; (f, v_h), \qquad \forall v_h \in V_h\,. \tag{1.4}$$

In the space $H_0^1(\Omega)$ the semi-norm $|u|_1 = \sqrt{(u', u')}$ becomes a norm due to Poincaré inequality. Furthermore, we define by $i_h u$ the finite element interpolant of the weak solution $u \in V := H_0^1(\Omega)$. Employing the Galerkin orthogonality, one can easily show that the error between the finite element solution and the finite element interpolant satisfies

$$|i_h u - u_h|_1 = \sup_{v_h \in V_h} \frac{\big((i_h u - u)', v_h'\big)}{|v_h|_1}\,. \tag{1.5}$$

Employing Cauchy–Schwarz inequality and interpolation estimate, we obtain for $u \in H^2(\Omega)$ the estimates

$$\big((i_h^1 u - u)', v_h'\big) \le |i_h^1 u - u|_1\, |v_h|_1 \le Ch|u|_1 |v_h|_1$$

and due to (1.5)

$$|i_h^1 u - u_h|_1 \leq Ch|u_h|_2 \,. \tag{1.6}$$

Now, we improve the bound of the term $\big((i_h u - u)', v_h'\big)$ using the nodal interpolation operator $i_h^1 : V \to V_h$ with

$$\begin{aligned} i_h^1 u(x_i) &= u(x_i), \quad i = 1, \ldots, M-1, \\ u(x_0) &= u(x_M) = 0 \,. \end{aligned} \tag{1.7}$$

For piecewise linear test functions $v_h \in V_h$ we get the following identity

$$\begin{aligned} \big((i_h^1 u - u)', v_h'\big) &= \sum_{i=0}^{M-1} \int_{x_i}^{x_{i+1}} (i_h^1 u - u)' v_h' dx \\ &= \sum_{i=0}^{M-1} \underbrace{\frac{v_h(x_{i+1}) - v_h(x_i)}{h}}_{=v_h'|_{(x_i, x_{i+1})}} \int_{x_i}^{x_{i+1}} (i_h^1 u - u)' dx \\ &= \sum_{i=0}^{M-1} \frac{v_h(x_{i+1}) - v_h(x_i)}{h} (i_h^1 u - u) \Big|_{x_i}^{x_{i+1}} \\ &= \sum_{i=0}^{M-1} \frac{v_h(x_{i+1}) - v_h(x_i)}{h} \Big\{ \big(i_h^1 u(x_{i+1}) - u(x_{i+1})\big) - \big(i_h^1 u(x_i) - u(x_i)\big) \Big\} \\ &= 0 \,, \end{aligned}$$

and from (1.5) we infer

$$|i_h^1 u - u_h|_1 = 0 \,. \tag{1.8}$$

Thus, the nodal interpolant $i_h^1 u$ coincides with the conforming piecewise linear discrete solution u_h, see Figure 1.1.

Using the fact that the error $|i_h^1 u - u_h|_1$ is superconvergent, we can post-process our discrete solution in order to obtain the higher accuracy. Let $\tau = (x_i, x_{i+2}) \in \mathcal{T}_{2h}$ be a macro–cell consisting of two child–cells $K_i = (x_i, x_{i+1})$ and $K_{i+1} = (x_{i+1}, x_{i+2})$, $i = 0, \ldots, M-1$, which originate from the regular refinement. We denote by $P_2(\tau)$ the space of quadratic polynomials over the macro–cell τ. Our aim is to construct a global post-processing operator

$$I_{2h}^2 : \; V \to S_{2h}^2 = \{v \in H_0^1(\Omega) : v|_\tau \in P_2(\tau)\}$$

which satisfies

$$I_{2h}^2 i_h^1 u = I_{2h}^2 u \quad \forall\, u \in V \,, \tag{1.9a}$$

$$|I_{2h}^2 u|_1 \leq C|u|_1 \quad \forall\, u \in V_h \,. \tag{1.9b}$$

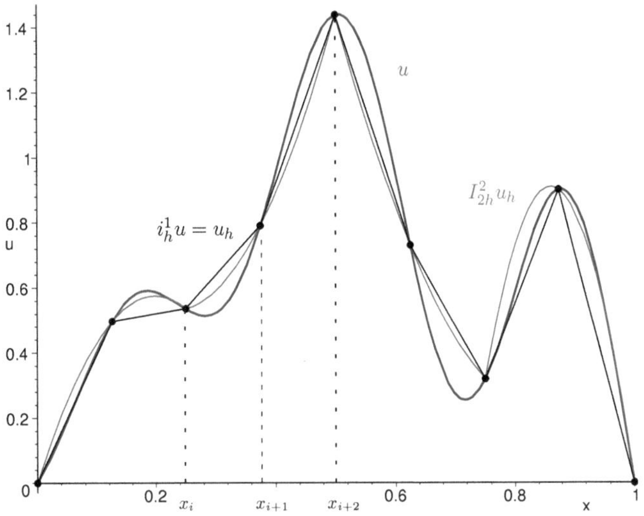

Figure 1.1: The piecewise linear conforming finite element solution u_h coincides with the nodal interpolant $i^1_h u$.

The post-processing interpolant can be defined on each $\tau \in \mathcal{T}_{2h}$ in the following way

$$\begin{gathered} I^2_{2h}u|_\tau \in P_2(\tau)\,, \\ I^2_{2h}u(x_i) = u(x_i)\,, \quad I^2_{2h}u(x_{i+1}) = u(x_{i+1})\,, \quad I^2_{2h}u(x_{i+2}) = u(x_{i+2})\,. \end{gathered} \tag{1.10}$$

The operator I^2_{2h} is then globally well–defined due to the continuity of u. The post-processing interpolator from (1.10) satisfies obviously (1.9a). Moreover, we have

$$I^2_{2h}u = u \quad \forall u \in P_2(\tau)\,.$$

Let $\hat\tau = (-1,1)$ and $F_\tau : \hat\tau \to \tau$ denote the reference cell and the affine reference mapping, respectively. Furthermore, we denote by $\widehat{V}$ the space of continuous, piecewise P_1 functions on τ. Setting $\hat u = u|_\tau \circ F_\tau$ and $(I_{2h}u)|_\tau = (\widehat{I}_{2h}\hat u) \circ F_\tau^{-1}$, it holds

$$|\widehat{I}_{2h}\hat u|_{1,\hat\tau} \le C|\hat u|_{1,\hat\tau} \qquad \forall\, \hat u \in \widehat{V}$$

since $|\widehat{I}_{2h} \cdot |_{1,\hat\tau}$ and $| \cdot |_{1,\hat\tau}$ are norms on the finite dimensional factor space $\widehat{V}/\mathbb{R}$ and since constant functions on $\hat\tau$ are exactly reproduced by the interpolator $\widehat{I}_{2h}$. Using scaling arguments, we obtain

$$|I^2_{2h}u|_{1,\tau} \le C|u|_{1,\tau} \quad \forall\, u \in V_h\,.$$

Property (1.9b) follows by summing up over all macro–cells τ. From Bramble–Hilbert lemma we obtain

$$|u - I_{2h}^2 u|_{1,\tau} \leq Ch^2 |u|_{3,\tau} ,$$

due to scaling arguments, and thus we get

$$|u - I_{2h}^2 u|_1 \leq Ch^2 |u|_3 \quad \forall u \in H^3(\Omega) . \tag{1.11}$$

From (1.9a) it follows

$$u - I_{2h}^2 u_h = u - I_{2h}^2 u + I_{2h}^2 (i_h^1 u - u_h) .$$

Then, the triangle inequality implies

$$|u - I_{2h}^2 u_h|_1 \leq |u - I_{2h}^2 u|_1 + |I_{2h}^2 (i_h^1 u - u_h)|_1 . \tag{1.12}$$

Collecting (1.11), (1.9b) and (1.8), we get from (1.12) the global superconvergence

$$|u - I_{2h}^2 u_h|_1 \leq Ch^2 |u|_3 . \tag{1.13}$$

The error between the post-processed solution and the weak solution $u \in H^3(\Omega)$ is with respect to the H^1 semi–norm one order better. Applying post-processing to the piecewise linear elements, we can obtain the accuracy of piecewise quadratic elements.

2 Reactor flow problem

2.1 Mathematical model

In this section we introduce the mathematical model describing incompressible isothermal flow without reaction. The considered equations for the velocity and pressure fields are related to those of fluid saturated porous media. Most of them use the Darcy law as a suitable model. However, there are restrictions of Darcy model, e.g. closely packed medium, flows at slow velocity. They can be circumvented with the Brinkman–Forchheimer-extended Darcy equation. Let $\Omega \subset \mathbb{R}^n$, $n = 2, 3$, represent the reactor channel. We denote its boundary by $\Gamma = \partial\Omega$. The conservation of volume-averaged values of momentum and mass in the packed reactor reads as follows

$$\begin{aligned} -\mathrm{div}\,(\varepsilon\nu\nabla\boldsymbol{u} - \varepsilon\boldsymbol{u}\otimes\boldsymbol{u}) + \frac{\varepsilon}{\varrho}\nabla p + \sigma(\boldsymbol{u}) &= \boldsymbol{f} \quad \text{in} \quad \Omega\,, \\ \mathrm{div}\,(\varepsilon\boldsymbol{u}) &= 0 \quad \text{in} \quad \Omega\,, \end{aligned} \tag{2.1}$$

where $\boldsymbol{u} : \Omega \to \mathbb{R}^n$, $p : \Omega \to \mathbb{R}$ denote the unknown velocity and pressure, respectively. The positive quantity $\varepsilon = \varepsilon(\boldsymbol{x})$ stands for porosity which describes the proportion of the non-solid volume to the total volume of material and varies spatially in general. The expression $\sigma(\boldsymbol{u})$ represents the friction forces caused by the packing and will be specified later on. The right-hand side $\boldsymbol{f}$ represents an outer force (e.g. gravitation), ϱ the constant fluid density and ν the constant kinematic viscosity of the fluid, respectively. The expression $\boldsymbol{u}\otimes\boldsymbol{u}$ symbolises the dyadic product of $\boldsymbol{u}$ with itself.

The formula given by Ergun [22] will be used to model the influence of the packing on the flow inertia effects

$$\sigma(\boldsymbol{u}) = 150\nu\frac{(1-\varepsilon)^2}{\varepsilon^2 d_p^2}\boldsymbol{u} + 1.75\frac{1-\varepsilon}{\varepsilon d_p}\boldsymbol{u}|\boldsymbol{u}|\,. \tag{2.2}$$

Thereby d_p stands for the diameter of pellets and $|\cdot|$ denotes the Euclidian vector norm. The linear term in (2.2) accounts for the head loss according to Darcy and the quadratic term according to Forchheimer law, respectively. For the derivation of the equations, modelling and homogenisation questions in porous media we refer to e.g. [9, 38]. To close the system (2.1) we prescribe Dirichlet boundary condition

$$\boldsymbol{u}|_\Gamma = \boldsymbol{g}\,, \tag{2.3}$$

whereby

$$\int_{\Gamma_i} \varepsilon \boldsymbol{g} \cdot \boldsymbol{n} \, ds = 0 \tag{2.4}$$

has to be fulfilled on each connected component Γ_i of the boundary Γ. We remark that in the case of polygonally bounded domain the outer normal vector $\boldsymbol{n}$ has jumps and thus the above integral should be replaced by a sum of integrals over each side of Γ.

The distribution of porosity ε is assumed to satisfy the following bounds

$$0 < \varepsilon_0 \le \varepsilon(\boldsymbol{x}) \le \varepsilon_1 \le 1 \quad \forall \boldsymbol{x} \in \Omega\,, \tag{A1}$$

with some constants $0 < \varepsilon_0,\ \varepsilon_1 \le 1$. In the next section we use the porosity distribution which is estimated for packed beds consisting of spherical particles and takes the near wall channelling effect into account. This kind of porosity distribution obeys assumption (A1).

Let us introduce dimensionless quantities

$$\boldsymbol{u}^* = \frac{\boldsymbol{u}}{U_0}\,, \quad p^* = \frac{p}{\varrho U_0^2}\,, \quad \boldsymbol{x}^* = \frac{\boldsymbol{x}}{d_p}\,, \quad \boldsymbol{g}^* = \frac{\boldsymbol{g}}{U_0}\,,$$

whereby U_0 denotes the magnitude of some reference velocity. For simplicity of notation we omit the asterisks. Then, the reactor flow problem reads in dimensionless form as follows

$$\left\{ \begin{array}{rcll} -\mathrm{div}\left(\frac{\varepsilon}{Re}\nabla \boldsymbol{u} - \varepsilon \boldsymbol{u} \otimes \boldsymbol{u}\right) + \varepsilon \nabla p + \frac{\alpha}{Re}\boldsymbol{u} + \beta \boldsymbol{u}|\boldsymbol{u}| & = & \boldsymbol{f} & \text{in } \Omega\,, \\ \mathrm{div}\,(\varepsilon \boldsymbol{u}) & = & 0 & \text{in } \Omega\,, \\ \boldsymbol{u} & = & \boldsymbol{g} & \text{on } \Gamma\,, \end{array} \right. \tag{2.5}$$

where

$$\alpha(\boldsymbol{x}) = 150\kappa^2(\boldsymbol{x})\,, \qquad \beta(\boldsymbol{x}) = 1.75\kappa(\boldsymbol{x}) \tag{2.6}$$

with

$$\kappa(\boldsymbol{x}) = \frac{1 - \varepsilon(\boldsymbol{x})}{\varepsilon(\boldsymbol{x})}\,, \tag{2.7}$$

and the Reynolds number is defined by

$$Re = \frac{U_0\, d_p}{\nu}\,.$$

The existence and uniqueness of solution of nonlinear model (2.5) with the constant porosity and without the convective term has been established in [17].

Remark 2.1 (2.5) *becomes a Navier-Stokes problem if* $\varepsilon \equiv 1$.

2.2 Existence and uniqueness results

In the following the porosity ε is assumed to belong to $W^{1,3}(\Omega) \cap L^\infty(\Omega)$. We start with the weak formulation of problem (2.5) and look for its solution in suitable Sobolev spaces.

2.2.1 Variational formulation

Let

$$L_0^2(\Omega) := \{v \in L^2(\Omega) : (v, 1) = 0\}$$

be the space consisting of L^2 functions with zero mean value. We define the spaces

$$\boldsymbol{X} := \boldsymbol{H}^1(\Omega), \quad \boldsymbol{X}_0 := \boldsymbol{H}_0^1(\Omega), \quad Q := L^2(\Omega), \quad M := L_0^2(\Omega),$$

and

$$\boldsymbol{V} := \boldsymbol{X}_0 \times M.$$

Let us introduce the following bilinear forms

$$a : \boldsymbol{X} \times \boldsymbol{X} \to \mathbb{R}, \qquad a(\boldsymbol{u}, \boldsymbol{v}) = \frac{1}{Re}\big(\varepsilon \nabla \boldsymbol{u}, \nabla \boldsymbol{v}\big),$$

$$b : \boldsymbol{X} \times Q \to \mathbb{R}, \qquad b(\boldsymbol{u}, q) = \big(\operatorname{div}(\varepsilon \boldsymbol{u}), q\big),$$

$$c : \boldsymbol{X} \times \boldsymbol{X} \to \mathbb{R}, \qquad c(\boldsymbol{u}, \boldsymbol{v}) = \frac{1}{Re}\big(\alpha \boldsymbol{u}, \boldsymbol{v}\big).$$

Furthermore, we define the semilinear form

$$d : \boldsymbol{X} \times \boldsymbol{X} \times \boldsymbol{X} \to \mathbb{R}, \qquad d(\boldsymbol{w}; \boldsymbol{u}, \boldsymbol{v}) = \big(\beta |\boldsymbol{w}| \boldsymbol{u}, \boldsymbol{v}\big),$$

and trilinearform

$$n : \boldsymbol{X} \times \boldsymbol{X} \times \boldsymbol{X} \to \mathbb{R}, \qquad n(\boldsymbol{w}, \boldsymbol{u}, \boldsymbol{v}) = \big((\varepsilon \boldsymbol{w} \cdot \nabla)\boldsymbol{u}, \boldsymbol{v}\big).$$

We set

$$A(\boldsymbol{w}; \boldsymbol{u}, \boldsymbol{v}) := a(\boldsymbol{u}, \boldsymbol{v}) + c(\boldsymbol{u}, \boldsymbol{v}) + n(\boldsymbol{w}, \boldsymbol{u}, \boldsymbol{v}) + d(\boldsymbol{w}; \boldsymbol{u}, \boldsymbol{v}).$$

Multiplying momentum and mass balances in (2.5) by test functions $\boldsymbol{v} \in \boldsymbol{X}_0$ and $q \in M$, respectively, and integrating by parts implies the weak formulation:

Find $(\boldsymbol{u}, p) \in \boldsymbol{X} \times M$ with $\boldsymbol{u}|_\Gamma = \boldsymbol{g}$ such that

$$A(\boldsymbol{u}; \boldsymbol{u}, \boldsymbol{v}) \quad b(\boldsymbol{v}, p) + b(\boldsymbol{u}, q) - (\boldsymbol{f}, \boldsymbol{v}) \quad \forall\, (\boldsymbol{v}, q) \in \boldsymbol{V}. \tag{2.8}$$

First, we recall the following result from [8]:

Theorem 2.2 *The mapping $u \mapsto \varepsilon u$ is an isomorphism from $H^1(\Omega)$ onto itself and from $H_0^1(\Omega)$ onto itself. It holds for all $u \in H^1(\Omega)$*

$$\|\varepsilon u\|_1 \le C\{\varepsilon_1 + |\varepsilon|_{1,3}\}\,\|u\|_1 \qquad \text{and} \qquad \left\|\frac{u}{\varepsilon}\right\|_1 \le C\left\{\varepsilon_0^{-1} + \varepsilon_0^{-2}\,|\varepsilon|_{1,3}\right\}\|u\|_1\,.$$

In the following the closed subspace of $\boldsymbol{H}_0^1(\Omega)$ defined by

$$\boldsymbol{W} = \{\boldsymbol{w} \in \boldsymbol{H}_0^1(\Omega): \quad b(\boldsymbol{w}, q) = 0 \quad \forall\, q \in L_0^2(\Omega)\}.$$

will be employed. Next, we establish and prove some properties of trilinear form $n(\cdot,\cdot,\cdot)$ and nonlinear form $d(\cdot;\cdot,\cdot)$.

Lemma 2.3 *Let $\boldsymbol{u}, \boldsymbol{v} \in \boldsymbol{H}^1(\Omega)$ and let $\boldsymbol{w} \in \boldsymbol{H}^1(\Omega)$ with $div(\varepsilon\boldsymbol{w}) = 0$ and $\boldsymbol{w}\cdot\boldsymbol{n}|_\Gamma = 0$. Then we have*

$$n(\boldsymbol{w},\boldsymbol{u},\boldsymbol{v}) = -n(\boldsymbol{w},\boldsymbol{v},\boldsymbol{u})\,. \tag{2.9}$$

Furthermore, the trilinear form $n(\cdot,\cdot,\cdot)$ and the nonlinear form $d(\cdot;\cdot,\cdot)$ are continuous, i.e.

$$|n(\boldsymbol{u},\boldsymbol{v},\boldsymbol{w})| \le C_\varepsilon\,\|\boldsymbol{u}\|_1\|\boldsymbol{v}\|_1\|\boldsymbol{w}\|_1 \quad \forall\,\boldsymbol{u},\boldsymbol{v},\boldsymbol{w} \in \boldsymbol{H}^1(\Omega)\,, \tag{2.10}$$

$$|d(\boldsymbol{u},\boldsymbol{v},\boldsymbol{w})| \le C_\varepsilon\,\|\boldsymbol{u}\|_1\|\boldsymbol{v}\|_1\|\boldsymbol{w}\|_1 \quad \forall\,\boldsymbol{u},\boldsymbol{v},\boldsymbol{w} \in \boldsymbol{H}^1(\Omega)\,, \tag{2.11}$$

and for $\boldsymbol{u} \in \boldsymbol{W}$ and for a sequence $\boldsymbol{u}^k \in \boldsymbol{W}$ with $\lim_{k\to\infty}\|\boldsymbol{u}^k - \boldsymbol{u}\|_0 = 0$, we have also

$$\lim_{k\to\infty} n(\boldsymbol{u}^k,\boldsymbol{u}^k,\boldsymbol{v}) = n(\boldsymbol{u},\boldsymbol{u},\boldsymbol{v}) \quad \forall\,\boldsymbol{v} \in \boldsymbol{W}. \tag{2.12}$$

Proof. We follow the proof of [30, Lemma 2.1, §2, Chapter IV] and adapt it to the trilinear form

$$n(\boldsymbol{w},\boldsymbol{u},\boldsymbol{v}) = ((\varepsilon\boldsymbol{w}\cdot\nabla)\boldsymbol{u},\boldsymbol{v}) = \sum_{i,j=1}^n (\varepsilon w_j \partial_j u_i, v_i)$$

which has the weighting factor ε. Hereby, symbols with subscripts denote components of bold faced vectors, e.g. $\boldsymbol{u} = (u_i)_{i=1,\ldots,n}$. Let $\boldsymbol{u} \in \boldsymbol{H}^1$, $\boldsymbol{v} \in \boldsymbol{D}(\Omega)$ and $\boldsymbol{w} \in \boldsymbol{W}$. Integrating by parts and employing density argument, we obtain immediately (2.9)

$$\begin{aligned}\sum_{i,j=1}^n (\varepsilon w_j \partial_j u_i, v_i) &= -\sum_{i,j=1}^n \left(\partial_j(\varepsilon w_j v_i), u_i\right) + \sum_{i,j=1}^n \langle \varepsilon w_j n_j u_i, v_i\rangle \\ &= -\sum_{i,j=1}^n (\varepsilon w_j \partial_j v_i, u_i) - (\mathrm{div}\,(\varepsilon\boldsymbol{w})\boldsymbol{u},\boldsymbol{v}) + \langle(\varepsilon\boldsymbol{w}\cdot\boldsymbol{n})\boldsymbol{u},\boldsymbol{v}\rangle \\ &= -n(\boldsymbol{w},\boldsymbol{v},\boldsymbol{u}).\end{aligned}$$

From Sobolev embedding $H^1(\Omega) \hookrightarrow L^4(\Omega)$ (see [1]) and Hölder inequality follows

$$\left|(\varepsilon w_j \partial_j u_i, v_i)\right| \le |\varepsilon|_{0,\infty}\,\|w_j\|_{0,4}\,\|\partial_j u_i\|_0\,\|v_i\|_{0,4} \le C\,|\varepsilon|_{0,\infty}\,\|w_j\|_1\,|u_i|_1\,\|v_i\|_1\,,$$

and consequently the proof of (2.10) is completed. Since $\lim_{k\to\infty} \|u_i^k u_j^k - u_i u_j\|_{0,1} = 0$ and $\varepsilon \partial_j v_i \in L^\infty(\Omega)$, the continuity estimate (2.10) implies

$$\lim_{k\to\infty} n(\boldsymbol{u}^k, \boldsymbol{u}^k, \boldsymbol{v}) = -\lim_{k\to\infty} n(\boldsymbol{u}^k, \boldsymbol{v}, \boldsymbol{u}^k) = -\lim_{k\to\infty} \sum_{i,j=1}^{n} \left(\varepsilon u_j^k\, \partial_j v_i^k, u_i^k\right)$$

$$= -\sum_{i,j=1}^{n} \left(\varepsilon u_j \partial_j v_i, u_i\right) = -n(\boldsymbol{u}, \boldsymbol{v}, \boldsymbol{u}) = n(\boldsymbol{u}, \boldsymbol{u}, \boldsymbol{v})\,.$$

The continuity of $d(\cdot;\cdot,\cdot)$ follows from Hölder inequality and Sobolev embedding $H^1(\Omega) \hookrightarrow L^4(\Omega)$ (see [1])

$$|d(\boldsymbol{u}; \boldsymbol{v}, \boldsymbol{w})| \le |\beta|_\infty \, \|\boldsymbol{u}\|_{0,4} \, \|\boldsymbol{v}\|_{0,4} \, \|\boldsymbol{w}\|_0 \le C_\varepsilon \|\boldsymbol{u}\|_1 \, \|\boldsymbol{v}\|_1 \, \|\boldsymbol{w}\|_1\,.$$

□

In the next stage we care about difficulties caused by prescribing inhomogeneous Dirichlet boundary condition. Analogous difficulties are already encountered in the analysis of Navier–Stokes problem. We carry out the study of three dimensional case. The extension in two dimensions is constructed analogously. Since $\boldsymbol{g} \in \boldsymbol{H}^{1/2}(\Gamma)$, we can extend $\boldsymbol{g}$ inside of Ω in the form of

$$\boldsymbol{g} = \varepsilon^{-1} \operatorname{curl} \boldsymbol{h}$$

with some $\boldsymbol{h} \in \boldsymbol{H}^2(\Omega)$. The operator curl is defined then as

$$\operatorname{curl} \boldsymbol{h} = (\partial_2 h_3 - \partial_3 h_2,\ \partial_3 h_1 - \partial_1 h_3,\ \partial_1 h_2 - \partial_2 h_1)\,.$$

We note that in the two dimensional case the vector potential $\boldsymbol{h} \in \boldsymbol{H}^2(\Omega)$ should be replaced by a scalar function $h \in H^2(\Omega)$ and the operator curl is then redefined as $\operatorname{curl} h = (\partial_2 h, -\partial_1 h)$. Our aim is to adapt the extension of Hopf (see [37]) to our model. We recall that for any parameter $\mu > 0$ there exists a scalar function $\varphi_\mu \in C^2(\bar{\Omega})$ such that

$$\left.\begin{array}{l}
\bullet\quad \varphi_\mu = 1 \text{ in some neighbourhood of } \Gamma \text{ (depending on } \mu)\,, \\
\bullet\quad \varphi_\mu(\boldsymbol{x}) = 0 \text{ if } d_\Gamma(\boldsymbol{x}) \ge 2\exp(-1/\mu)\,, \text{ where } d_\Gamma(\boldsymbol{x}) := \inf_{\boldsymbol{y}\in\Gamma} |\boldsymbol{x} - \boldsymbol{y}| \\
\qquad \text{denotes the distance of } \boldsymbol{x} \text{ to } \Gamma\,, \\
\bullet\quad |\partial_j \varphi_\mu(\boldsymbol{x})| \le \mu / d_\Gamma(\boldsymbol{x}) \ \text{ if } \ d_\Gamma(\boldsymbol{x}) < 2\exp(-1/\mu)\,,\ j = 1, \ldots, n\,.
\end{array}\right\} \quad \text{(Ex)}$$

For the construction of φ_μ see also [30, Lemma 2.4, §2, Chapter IV].

Let us define

$$\boldsymbol{g}_\mu := \varepsilon^{-1} \operatorname{curl} (\varphi_\mu \boldsymbol{h}) \tag{2.13}$$

In the following lemma we establish bounds which are crucial for proving existence of velocity.

Lemma 2.4 *The function $\boldsymbol{g}_\mu$ satisfies the following conditions*

$$div(\varepsilon \boldsymbol{g}_\mu) = 0, \quad \boldsymbol{g}_\mu|_\Gamma = \boldsymbol{g} \qquad \forall\, \mu > 0\,, \tag{2.14}$$

and for any $\delta > 0$ there exists sufficiently small $\mu > 0$ such that

$$|d(\boldsymbol{u}+\boldsymbol{g}_\mu; \boldsymbol{g}_\mu, \boldsymbol{u})| \le \delta\, \|\beta\|_{0,\infty}\, |\boldsymbol{u}|_1 \big(|\boldsymbol{u}|_1 + \|\boldsymbol{g}_\mu\|_0\big) \qquad \forall\, \boldsymbol{u} \in \boldsymbol{X}_0\,, \tag{2.15}$$

$$|n(\boldsymbol{u}, \boldsymbol{g}_\mu, \boldsymbol{u})| \le \delta\, |\boldsymbol{u}|_1^2 \qquad \forall\, \boldsymbol{u} \in \boldsymbol{W}\,. \tag{2.16}$$

Proof. The relations in (2.14) are obvious. We follow [17] in order to show (2.15). Since $\boldsymbol{h} \in \boldsymbol{H}^2(\Omega)$ Sobolev's embedding theorem implies $\boldsymbol{h} \in \boldsymbol{L}^\infty(\Omega)$, so we get according to the properties of φ_μ in (Ex) the following bound

$$|\boldsymbol{g}_\mu| \le C\,\varepsilon_0^{-1} \left\{ |\nabla \boldsymbol{h}| + \frac{\mu}{d_\Gamma(\boldsymbol{x})} |\boldsymbol{h}| \right\} \le C \left\{ \frac{\mu}{d_\Gamma(\boldsymbol{x})} + |\nabla \boldsymbol{h}| \right\}.$$

Defining

$$\Omega_\mu := \{\boldsymbol{x} \in \Omega :\ d_\Gamma(\boldsymbol{x}) < 2\exp(-1/\mu)\}$$

we obtain from Cauchy-Schwarz and triangle inequalities

$$|(\beta|\boldsymbol{u}+\boldsymbol{g}_\mu|, \boldsymbol{g}_\mu \cdot \boldsymbol{u})| \le \|\beta\|_{0,\infty}\, \|\boldsymbol{u}\|_0\, \|\boldsymbol{u}\cdot\boldsymbol{g}_\mu\|_{0,\Omega_\mu} + \|\beta\|_{0,\infty}\, \|\boldsymbol{g}_\mu\|_0\, \|\boldsymbol{u}\cdot\boldsymbol{g}_\mu\|_{0,\Omega_\mu}\,, \tag{2.17}$$

$$\begin{aligned}
\|\boldsymbol{u}\cdot\boldsymbol{g}_\mu\|_{0,\Omega_\mu}^2 &\le \int_{\Omega_\mu} |\boldsymbol{u}|^2 |\boldsymbol{g}_\mu|^2 d\boldsymbol{x} \\
&\le C \int_{\Omega_\mu} |\boldsymbol{u}|^2 \Big\{ \big(\mu/d_\Gamma(\boldsymbol{x})\big)^2 + 2\mu/d_\Gamma(\boldsymbol{x})\, |\nabla \boldsymbol{h}| + |\nabla \boldsymbol{h}|^2 \Big\} d\boldsymbol{x} \\
&\le C \Big\{ \mu^2 \|\boldsymbol{u}/d_\Gamma\|_{0,\Omega_\mu}^2 + 2\mu \|\boldsymbol{u}/d_\Gamma\|_{0,\Omega_\mu} \|\boldsymbol{u}\|_{0,4,\Omega_\mu} \big\||\nabla \boldsymbol{h}|\big\|_{0,4,\Omega_\mu} + \|\boldsymbol{u}\|_{0,4,\Omega_\mu}^2 \big\||\nabla \boldsymbol{h}|\big\|_{0,4,\Omega_\mu}^2 \Big\} \\
&\le C \Big\{ \mu \|\boldsymbol{u}/d_\Gamma\|_{0,\Omega_\mu} + \|\boldsymbol{u}\|_{0,4} \big\||\nabla \boldsymbol{h}|\big\|_{0,4,\Omega_\mu} \Big\}^2,
\end{aligned}$$

and consequently

$$\|\boldsymbol{u}\cdot\boldsymbol{g}_\mu\|_{0,\Omega_\mu} \le C \Big\{ \mu \|\boldsymbol{u}/d_\Gamma\|_{0,\Omega_\mu} + \|\boldsymbol{u}\|_{0,4} \big\||\nabla \boldsymbol{h}|\big\|_{0,4,\Omega_\mu} \Big\}\,. \tag{2.18}$$

Applying Hardy inequality (see [1])

$$\|v/d_\Gamma\|_0 \le C|v|_1 \quad \forall\, v \in H_0^1(\Omega)$$

and using Sobolev embedding $H^1(\Omega) \hookrightarrow L^4(\Omega)$, estimate (2.18) becomes

$$\|\boldsymbol{u}\cdot\boldsymbol{g}_\mu\|_{0,\Omega_\mu} \le C\lambda(\mu) \|\boldsymbol{u}\|_1, \tag{2.19}$$

where

$$\lambda(\mu) := \max\big\{\mu,\ \big\||\nabla \boldsymbol{h}|\big\|_{0,4,\Omega_\mu}\big\}\,.$$

From (2.17), (2.19), Poincaré inequality and from the fact that $\lim_{\mu\to 0} \lambda(\mu) = 0$ we conclude that for any $\delta > 0$ we can choose sufficiently small $\mu > 0$ such that

$$|(\beta\,|\boldsymbol{u}+\boldsymbol{g}_\mu|\boldsymbol{g}_\mu, \boldsymbol{u})| \le \delta\, \|\beta\|_{0,\infty}\, |\boldsymbol{u}|_1 \big(|\boldsymbol{u}|_1 + \|\boldsymbol{g}_\mu\|_0\big)$$

holds. Therefore the proof of estimate (2.15) is completed. Now, we take a look at the trilinear convective term

$$\begin{aligned} n(\boldsymbol{u}, \boldsymbol{g}_\mu, \boldsymbol{u}) &= ((\varepsilon \boldsymbol{u}\cdot\nabla)\boldsymbol{g}_\mu, \boldsymbol{u})_{\Omega_\mu} = \Big((\varepsilon \boldsymbol{u}\cdot\nabla)\left\{\varepsilon^{-1}\operatorname{curl}(\varphi_\mu \boldsymbol{h})\right\}, \boldsymbol{u}\Big)_{\Omega_\mu} \\ &= \Big((\boldsymbol{u}\cdot\nabla)\left\{\operatorname{curl}(\varphi_\mu \boldsymbol{h})\right\}, \boldsymbol{u}\Big)_{\Omega_\mu} - ((\boldsymbol{u}\cdot\nabla\varepsilon)\,\boldsymbol{g}_\mu, \boldsymbol{u})_{\Omega_\mu}\,. \end{aligned}$$

The first term of above difference becomes small due to [30, Lemma 2.3, §2, Chapter IV]

$$\left|((\boldsymbol{u}\cdot\nabla)\left\{\operatorname{curl}(\varphi_\mu \boldsymbol{h})\right\}, \boldsymbol{u})_{\Omega_\mu}\right| = \left|((\boldsymbol{u}\cdot\nabla)(\varepsilon \boldsymbol{g}_\mu), \boldsymbol{u})_{\Omega_\mu}\right| \le \delta |\boldsymbol{u}|_1^2 \tag{2.20}$$

as long as $\mu > 0$ is chosen sufficiently small. Using Hölder inequality, Sobolev embedding $H^1(\Omega) \hookrightarrow L^6(\Omega)$ yields

$$\left|((\boldsymbol{u}\cdot\nabla\varepsilon)\,\boldsymbol{g}_\mu, \boldsymbol{u})_{\Omega_\mu}\right| \le C \|\varepsilon\|_{1,3}\, \|\boldsymbol{g}_\mu\cdot\boldsymbol{u}\|_0\, \|\boldsymbol{u}\|_1$$

which together with (2.19) implies for sufficiently small $\mu > 0$ the bound

$$\left|((\boldsymbol{u}\cdot\nabla\varepsilon)\,\boldsymbol{g}_\mu, \boldsymbol{u})_{\Omega_\mu}\right| \le \delta |\boldsymbol{u}|_1^2\,. \tag{2.21}$$

From (2.20) and (2.21) follows the desired estimate (2.16). □

While the general framework for linear and non-symmetric saddle point problems can be found in [8], our problem requires more attention due to its nonlinear character. Setting $\boldsymbol{w} := \boldsymbol{u} - \boldsymbol{g}_\mu$, the weak formulation (2.8) is equivalent to the following problem

Find $(\boldsymbol{w}, p) \in \boldsymbol{V}$ such that

$$A(\boldsymbol{w}+\boldsymbol{g}_\mu; \boldsymbol{w}+\boldsymbol{g}_\mu, \boldsymbol{v}) - b(\boldsymbol{v}, p) + b(\boldsymbol{w}+\boldsymbol{g}_\mu, q) = (\boldsymbol{f}, \boldsymbol{v}) \quad \forall\, (\boldsymbol{v}, q) \in \boldsymbol{V}\,. \tag{2.22}$$

Let us define the nonlinear mapping $G:\ \boldsymbol{W} \to \boldsymbol{W}$ with

$$\begin{aligned} \big[G(\boldsymbol{w}), \boldsymbol{v}\big] :=& a(\boldsymbol{w}+\boldsymbol{g}_\mu, \boldsymbol{v}) + c(\boldsymbol{w}+\boldsymbol{g}_\mu, \boldsymbol{v}) - (\boldsymbol{f}, \boldsymbol{v}) \\ &+ n(\boldsymbol{w}+\boldsymbol{g}_\mu, \boldsymbol{w}+\boldsymbol{g}_\mu, \boldsymbol{v}) + d(\boldsymbol{w}+\boldsymbol{g}_\mu; \boldsymbol{w}+\boldsymbol{g}_\mu, \boldsymbol{v})\,, \end{aligned} \tag{2.23}$$

whereby $[\cdot,\cdot]$ defines the inner product in $\boldsymbol{W}$ via $[u, v] := (\nabla u, \nabla v)$. Then, the variational problem (2.22) reads in the space of ε-weighted divergence free functions $\boldsymbol{W}$ as follows

Find $\boldsymbol{w} \in \boldsymbol{W}$ such that

$$\big[G(\boldsymbol{w}), \boldsymbol{v}\big] = 0 \quad \forall\, \boldsymbol{v} \in \boldsymbol{W}. \tag{2.24}$$

2.2.2 Solvability of nonlinear saddle point problem

We start our study of nonlinear operator problem (2.24) with the following lemma.

Lemma 2.5 *The mapping G defined in (2.23) is continuous and there exists $r > 0$ such that*

$$\big[G(\boldsymbol{u}), \boldsymbol{u}\big] > 0 \quad \forall\, \boldsymbol{u} \in \boldsymbol{W} \quad \text{with} \quad |\boldsymbol{u}|_1 = r. \tag{2.25}$$

Proof. Let $(\boldsymbol{u}^k)_{k\in\mathbf{N}}$ be a sequence in $\boldsymbol{W}$ with $\lim_{k\to\infty} \|\boldsymbol{u}^k - \boldsymbol{u}\|_1 = 0$. Then, applying Cauchy–Schwarz inequality and (2.16), we obtain for any $\boldsymbol{v} \in \boldsymbol{W}$

$$\begin{aligned}
\big|\big[G(\boldsymbol{u}^k) - G(\boldsymbol{u}), \boldsymbol{v}\big]\big| &\le \frac{1}{Re}\big|\big(\varepsilon\nabla(\boldsymbol{u}^k - \boldsymbol{u}), \nabla\boldsymbol{v}\big)\big| + \frac{1}{Re}\big|\big(\alpha(\boldsymbol{u}^k - \boldsymbol{u}), \boldsymbol{v}\big)\big| \\
&\quad + \big|\big(\beta|\boldsymbol{u}^k + \boldsymbol{g}_\mu|(\boldsymbol{u}^k - \boldsymbol{u}), \boldsymbol{v}\big)\big| + \big|\big(\beta(|\boldsymbol{u}^k + \boldsymbol{g}_\mu| - |\boldsymbol{u} + \boldsymbol{g}_\mu|)(\boldsymbol{u} + \boldsymbol{g}_\mu), \boldsymbol{v}\big)\big| \\
&\quad + \big|n(\boldsymbol{u}^k, \boldsymbol{u}^k, \boldsymbol{v}) - n(\boldsymbol{u}, \boldsymbol{u}, \boldsymbol{v})\big| + \big|n(\boldsymbol{u}^k - \boldsymbol{u}, \boldsymbol{g}_\mu, \boldsymbol{v})\big| + \big|n(\boldsymbol{g}_\mu, \boldsymbol{u}^k - \boldsymbol{u}, , \boldsymbol{v})\big| \\
&\le \frac{\varepsilon_1}{Re}|\boldsymbol{u}^k - \boldsymbol{u}|_1|\boldsymbol{v}|_1 + \frac{1}{Re}\|\alpha\|_{0,\infty}\|\boldsymbol{u}^k - \boldsymbol{u}\|_0\|\boldsymbol{v}\|_0 \\
&\quad + \|\beta\|_{0,\infty}\|\boldsymbol{u}^k + \boldsymbol{g}_\mu\|_{0,4}\|\boldsymbol{u}^k - \boldsymbol{u}\|_0\|\boldsymbol{v}\|_{0,4} + \|\beta\|_{0,\infty}\|\boldsymbol{u} + \boldsymbol{g}_\mu\|_{0,4}\|\boldsymbol{u}^k - \boldsymbol{u}\|_0\|\boldsymbol{v}\|_{0,4} \\
&\quad + \big|n(\boldsymbol{u}^k, \boldsymbol{u}^k, \boldsymbol{v}) - n(\boldsymbol{u}, \boldsymbol{u}, \boldsymbol{v})\big| + C\|\boldsymbol{u}^k - \boldsymbol{u}\|_1\|\boldsymbol{g}_\mu\|_1\|\boldsymbol{v}\|_1 .
\end{aligned}$$

The boundedness of $\boldsymbol{u}^k$ in $\boldsymbol{W}$, (2.12) and Poincaré inequality imply

$$\big|\big[G(\boldsymbol{u}^k) - G(\boldsymbol{u}), \boldsymbol{v}\big]\big| \to 0 \quad \text{as} \quad k \to \infty \qquad \forall\, \boldsymbol{v} \in \boldsymbol{W} .$$

Thus, employing

$$|G(\boldsymbol{u}^k) - G(\boldsymbol{u})|_1 = \sup_{\substack{\boldsymbol{v}\in\boldsymbol{W}\\ \boldsymbol{v}\neq 0}} \frac{\big[G(\boldsymbol{u}^k) - G(\boldsymbol{u}), \boldsymbol{v}\big]}{|\boldsymbol{v}|_1},$$

we state that G is continuous. Now, we note that for any $\boldsymbol{u} \in \boldsymbol{W}$ we have

$$\begin{aligned}
\big[G(\boldsymbol{u}), \boldsymbol{u}\big] &= \frac{1}{Re}(\varepsilon\nabla(\boldsymbol{u} + \boldsymbol{g}_\mu), \nabla\boldsymbol{u}) + \frac{1}{Re}(\alpha(\boldsymbol{u} + \boldsymbol{g}_\mu), \boldsymbol{u}) \\
&\quad + \big(\beta|\boldsymbol{u} + \boldsymbol{g}_\mu|(\boldsymbol{u} + \boldsymbol{g}_\mu), \boldsymbol{u}\big) + n(\boldsymbol{u} + \boldsymbol{g}_\mu, \boldsymbol{u} + \boldsymbol{g}_\mu, \boldsymbol{u}) - (\boldsymbol{f}, \boldsymbol{u}) \\
&\ge \frac{\varepsilon_0}{Re}|\boldsymbol{u}|_1^2 - \frac{\varepsilon_1}{Re}|(\nabla\boldsymbol{g}_\mu, \nabla\boldsymbol{u})| + \frac{1}{Re}(\alpha\boldsymbol{u}, \boldsymbol{u}) - \frac{1}{Re}|(\alpha\boldsymbol{g}_\mu, \boldsymbol{u})| \\
&\quad + (\beta|\boldsymbol{u} + \boldsymbol{g}_\mu|, |\boldsymbol{u}|^2) - \big|(\beta|\boldsymbol{u} + \boldsymbol{g}_\mu|\boldsymbol{g}_\mu, \boldsymbol{u})\big| \\
&\quad + n(\boldsymbol{u}, \boldsymbol{g}_\mu, \boldsymbol{u}) + n(\boldsymbol{g}_\mu, \boldsymbol{g}_\mu, \boldsymbol{u}) - \|\boldsymbol{f}\|_0\|\boldsymbol{u}\|_0 \\
&\ge \frac{\varepsilon_0}{Re}|\boldsymbol{u}|_1^2 - \frac{\varepsilon_1}{Re}|\boldsymbol{g}_\mu|_1|\boldsymbol{u}|_1 \\
&\quad - \frac{1}{Re}\|\alpha\|_{0,\infty}\|\boldsymbol{g}_\mu\|_0\|\boldsymbol{u}\|_0 - \big|(\beta|\boldsymbol{u} + \boldsymbol{g}_\mu|\boldsymbol{g}_\mu, \boldsymbol{u})\big| \\
&\quad - \big|n(\boldsymbol{u}, \boldsymbol{g}_\mu, \boldsymbol{u})\big| - C\|\boldsymbol{g}_\mu\|_1^2\|\boldsymbol{u}\|_1 - \|\boldsymbol{f}\|_0\|\boldsymbol{u}\|_0 .
\end{aligned} \tag{2.26}$$

From Poincaré inequality we infer the estimate

$$\|v\|_1 \le C|v|_1 \quad \forall\, v \in H_0^1(\Omega),$$

which together with (2.15), (2.16) and (2.26) results in

$$\begin{aligned}[G(\boldsymbol{u}), \boldsymbol{u}] \ge & \left\{\frac{\varepsilon_0}{Re} - \delta(1 + \|\beta\|_{0,\infty})\right\} |\boldsymbol{u}|_1^2 \\ & - \left\{\frac{\varepsilon_1}{Re}|\boldsymbol{g}_\mu|_1 + C_1\frac{1}{Re}\|\alpha\|_{0,\infty}\|\boldsymbol{g}_\mu\|_0 + \delta\|\beta\|_{0,\infty}\|\boldsymbol{g}_\mu\|_0 + C_2\|\boldsymbol{g}_\mu\|_1^2 + C_3\|\boldsymbol{f}\|_0\right\}|\boldsymbol{u}|_1.\end{aligned}$$

Choosing δ such that

$$0 < \delta < \delta_0 := \frac{\varepsilon_0}{Re}\left(1 + \|\beta\|_{0,\infty}\right)^{-1},$$

and $r > r_0$ with

$$r_0 := \frac{\frac{\varepsilon_1}{Re}|\boldsymbol{g}_\mu|_1 + \frac{1}{Re}C_1\|\alpha\|_{0,\infty}\|\boldsymbol{g}_\mu\|_0 + \delta\|\beta\|_{0,\infty}\|\boldsymbol{g}_\mu\|_0 + C_2\|\boldsymbol{g}_\mu\|_1^2 + C_3\|\boldsymbol{f}\|_0}{\frac{\varepsilon_0}{Re} - \delta(1 + \|\beta\|_{0,\infty})}, \tag{2.27}$$

leads to the desired assertion (2.25). □

The following lemma plays a key role in the existence proof.

Lemma 2.6 *Let Y be finite-dimensional Hilbert space with inner product $[\cdot,\cdot]$ inducing a norm $\|\cdot\|$, and $T : Y \to Y$ be a continuous mapping such that*

$$[T(x), x] > 0 \quad \text{for} \quad \|x\| = r_0 > 0.$$

Then there exists $x \in Y$, with $\|x\| \le r_0$, such that

$$T(x) = 0.$$

Proof. See [59]. □

Now we are able to prove the main result concerning existence of velocity.

Theorem 2.7 *The problem (2.24) has at least one solution $\boldsymbol{u} \in \boldsymbol{W}$.*

Proof. We construct the approximate sequence of Galerkin solutions. Since the space $\boldsymbol{W}$ is separable, there exists a sequence of linearly independent elements $(\boldsymbol{w}^i)_{i\in\mathbb{N}} \subset \boldsymbol{W}$. Let $\boldsymbol{X}_m$ be the finite dimensional subspace of $\boldsymbol{W}$ with

$$\boldsymbol{X}_m := \operatorname{span}\{\boldsymbol{w}^i,\ i = 1, \ldots, m\}$$

and endowed with the scalar product of $\boldsymbol{W}$. Let $\boldsymbol{u}^m = \sum_{j=1}^{m} a_j \boldsymbol{w}^j$, $a_j \in \mathbb{R}$, be a Galerkin solution of (2.24) defined by

$$\left[G(\boldsymbol{u}^m), \boldsymbol{w}^j\right] = 0, \quad \forall\, j = 1, \ldots, m. \tag{2.28}$$

From Lemma 2.5 and Lemma 2.6 we conclude that

$$\left[G(\boldsymbol{u}^m), \boldsymbol{w}\right] = 0 \quad \forall\, \boldsymbol{w} \in \boldsymbol{X}_m \tag{2.29}$$

has a solution $\boldsymbol{u}^m \in \boldsymbol{X}_m$. The unknown coefficients a_j can be obtained from the algebraic system (2.28). On the other hand, multiplying (2.28) by a_j, and adding the equations for $j = 1, \ldots, m$ we have

$$\begin{aligned} 0 &= \left[G(\boldsymbol{u}^m), \boldsymbol{u}^m\right] \\ &\geq \left\{\frac{1}{Re} - \delta(1 + \|\beta\|_{0,\infty})\right\} |\boldsymbol{u}^m|_1^2 \\ &\quad - \left\{\frac{1}{Re}|\boldsymbol{g}_\mu|_1 + C_1\frac{1}{Re}\|\alpha\|_{0,\infty}\|\boldsymbol{g}_\mu\|_0 + \delta\|\beta\|_{0,\infty}\|\boldsymbol{g}_\mu\|_0 + C_2\|\boldsymbol{g}_\mu\|_1^2 + C_3\|\boldsymbol{f}\|_0\right\}|\boldsymbol{u}^m|_1. \end{aligned}$$

This gives together with (2.27) the uniform boundedness in $\boldsymbol{W}$

$$|\boldsymbol{u}^m|_1 \leq r_0,$$

therefore there exists $\boldsymbol{u} \in \boldsymbol{W}$ and a subsequence $m_k \longrightarrow \infty$ (we write for the convenience m instead of m_k) such that

$$\boldsymbol{u}^m \rightharpoonup \boldsymbol{u} \quad \text{in} \quad \boldsymbol{W}.$$

Furthermore, the compactness of embedding $H^1(\Omega) \hookrightarrow L^4(\Omega)$ implies

$$\boldsymbol{u}^m \longrightarrow \boldsymbol{u} \quad \text{in} \quad \boldsymbol{L}^4(\Omega).$$

Taking the limit in (2.29) with $m \longrightarrow \infty$ we get

$$\left[G(\boldsymbol{u}), \boldsymbol{w}\right] = 0 \quad \forall\, \boldsymbol{w} \in \boldsymbol{X}_m. \tag{2.30}$$

Finally, we apply the continuity argument and state that (2.30) is preserved for any $\boldsymbol{w} \in \boldsymbol{W}$, therefore $\boldsymbol{u}$ is the solution of (2.24). □

For the reconstruction of the pressure we need inf-sup-theorem

Theorem 2.8 *Assume that the bilinear form $b(\cdot,\cdot)$ satisfies the inf-sup condition*

$$\inf_{q\in M} \sup_{\boldsymbol{v}\in \boldsymbol{X}_0} \frac{b(\boldsymbol{v}, q)}{|\boldsymbol{v}|_1\, \|q\|_0} \geq \gamma > 0. \tag{2.31}$$

Then, for each solution $\boldsymbol{u}$ of the nonlinear problem (2.24) there exists a unique pressure $p \in M$ such that the pair $(\boldsymbol{u}, p) \in \boldsymbol{V}$ is a solution of the homogeneous problem (2.22).

Proof. See [30, Theorem 1.4, §1, Chapter IV]. □

We end up this subsection by constituting existence of the pressure.

Theorem 2.9 *Let $\boldsymbol{w}$ be solution of problem (2.24). Then, there exists unique pressure $p \in M$.*

Proof. We verify the inf-sup condition (2.31) of Theorem 2.8 by employing the isomorphism of Theorem 2.2. From [30, Corollary 2.4, §2, Chapter I] follows that for any q in $L^2_0(\Omega)$ there exists $\boldsymbol{v}$ in $\boldsymbol{H}^1_0(\Omega)$ such that

$$(\operatorname{div} \boldsymbol{v}, q) \geq \gamma^* \|\boldsymbol{v}\|_1 \|q\|_0$$

with a positive constant γ^*. Setting $\boldsymbol{u} = \boldsymbol{v}/\varepsilon$ and applying the isomorphism in Theorem 2.2, we obtain the estimate

$$b(\boldsymbol{u}, q) = (\operatorname{div} \boldsymbol{v}, q) \geq \gamma^* \|\boldsymbol{v}\|_1 \|q\|_0 \geq \gamma_\varepsilon \|\boldsymbol{u}\|_1 \|q\|_0$$

where $\gamma_\varepsilon = \dfrac{\gamma^*}{C\left\{\varepsilon_0^{-1} + \varepsilon_0^{-2} |\varepsilon|_{1,3}\right\}}$. From the above estimate we conclude the inf-sup condition (2.31). □

2.2.3 Uniqueness of weak solution

We exploit a priori estimates in order to prove uniqueness of weak velocity and pressure.

Theorem 2.10 *If $\|\boldsymbol{g}_\mu\|_1$, $\|\boldsymbol{f}\|_{-1} := \sup\limits_{0 \neq \boldsymbol{v} \in \boldsymbol{H}^1(\Omega)} \dfrac{(\boldsymbol{f}, \boldsymbol{v})}{\|\boldsymbol{v}\|_1}$ are sufficiently small, then the solution of (2.24) is unique.*

Proof. Assume that $(\boldsymbol{u}_1, p_1)$ and $(\boldsymbol{u}_2, p_2)$ are two different solutions of (2.22). From (2.9) in Lemma 2.3 we obtain $n(\boldsymbol{w}, \boldsymbol{u}, \boldsymbol{u}) = 0 \;\; \forall\, \boldsymbol{w}, \boldsymbol{u} \in \boldsymbol{W}$. Then, we obtain

$$
\begin{aligned}
0 &= \big[G(\boldsymbol{u}_1) - G(\boldsymbol{u}_2), \boldsymbol{u}_1 - \boldsymbol{u}_2\big] \\
&= a(\boldsymbol{u}_1 - \boldsymbol{u}_2, \boldsymbol{u}_1 - \boldsymbol{u}_2) + c(\boldsymbol{u}_1 - \boldsymbol{u}_2, \boldsymbol{u}_1 - \boldsymbol{u}_2) - (\boldsymbol{f}, \boldsymbol{u}_1 - \boldsymbol{u}_2) \\
&\quad + n(\boldsymbol{u}_1 + \boldsymbol{g}_\mu, \boldsymbol{u}_1 + \boldsymbol{g}_\mu, \boldsymbol{u}_1 - \boldsymbol{u}_2) - n(\boldsymbol{u}_2 + \boldsymbol{g}_\mu, \boldsymbol{u}_2 + \boldsymbol{g}_\mu, \boldsymbol{u}_1 - \boldsymbol{u}_2) \\
&\quad + (\beta|\boldsymbol{u}_1 + \boldsymbol{g}_\mu|(\boldsymbol{u}_1 + \boldsymbol{g}_\mu), \boldsymbol{u}_1 - \boldsymbol{u}_2) \\
&\quad - \big(\beta|\boldsymbol{u}_2 + \boldsymbol{g}_\mu|(\boldsymbol{u}_2 + \boldsymbol{g}_\mu), \boldsymbol{u}_1 - \boldsymbol{u}_2\big) \\
&\geq \frac{\varepsilon_0}{Re}|\boldsymbol{u}_1 - \boldsymbol{u}_2|_1^2 - \|\boldsymbol{f}\|_{-1}\|\boldsymbol{u}_1 - \boldsymbol{u}_2\|_1 \\
&\quad + n(\boldsymbol{u}_1 - \boldsymbol{u}_2, \boldsymbol{u}_2 + \boldsymbol{g}_\mu, \boldsymbol{u}_1 - \boldsymbol{u}_2) \\
&\quad + \big(\beta|\boldsymbol{u}_1 + \boldsymbol{g}_\mu|(\boldsymbol{u}_1 - \boldsymbol{u}_2), \boldsymbol{u}_1 - \boldsymbol{u}_2\big) \\
&\quad + \big(\beta(|\boldsymbol{u}_1 + \boldsymbol{g}_\mu| - |\boldsymbol{u}_2 + \boldsymbol{g}_\mu|)(\boldsymbol{u}_2 + \boldsymbol{g}_\mu), \boldsymbol{u}_1 - \boldsymbol{u}_2\big) \\
&\geq \frac{\varepsilon_0}{Re}|\boldsymbol{u}_1 - \boldsymbol{u}_2|_1^2 - \|\boldsymbol{f}\|_{-1}\|\boldsymbol{u}_1 - \boldsymbol{u}_2\|_1 \\
&\quad - |n(\boldsymbol{u}_1 - \boldsymbol{u}_2, \boldsymbol{u}_2, \boldsymbol{u}_1 - \boldsymbol{u}_2)| - \big|n(\boldsymbol{u}_1 - \boldsymbol{u}_2, \boldsymbol{g}_\mu, \boldsymbol{u}_1 - \boldsymbol{u}_2)\big| \\
&\quad - \|\beta\|_{0,\infty}\left|\big(|\boldsymbol{u}_1 + \boldsymbol{g}_\mu| \cdot |\boldsymbol{u}_1 - \boldsymbol{u}_2|, |\boldsymbol{u}_1 - \boldsymbol{u}_2|\big)\right| \\
&\quad - \|\beta\|_{0,\infty}\left|\big(\big||\boldsymbol{u}_1 + \boldsymbol{g}_\mu| - |\boldsymbol{u}_2 + \boldsymbol{g}_\mu|\big| \cdot |\boldsymbol{u}_2 + \boldsymbol{g}_\mu|, |\boldsymbol{u}_1 - \boldsymbol{u}_2|\big)\right| .
\end{aligned}
\tag{2.32}
$$

From Cauchy-Schwarz inequality and Sobolev embedding $H^1(\Omega) \hookrightarrow L^4(\Omega)$ we deduce

$$
\left|\big(|\boldsymbol{u}_1 + \boldsymbol{g}_\mu| \cdot |\boldsymbol{u}_1 - \boldsymbol{u}_2|, |\boldsymbol{u}_1 - \boldsymbol{u}_2|\big)\right| \leq C\left\{\|\boldsymbol{u}_1\|_0 + \|\boldsymbol{g}_\mu\|_0\right\}\|\boldsymbol{u}_1 - \boldsymbol{u}_2\|_1^2, \tag{2.33}
$$

$$
\begin{aligned}
&\left|\big(\big||\boldsymbol{u}_1 + \boldsymbol{g}_\mu| - |\boldsymbol{u}_2 + \boldsymbol{g}_\mu|\big| \cdot |\boldsymbol{u}_2 + \boldsymbol{g}_\mu|, |\boldsymbol{u}_1 - \boldsymbol{u}_2|\big)\right| \\
&\leq C\left\{\|\boldsymbol{u}_2\|_0 + \|\boldsymbol{g}_\mu\|_0\right\}\|\boldsymbol{u}_1 - \boldsymbol{u}_2\|_1^2,
\end{aligned}
\tag{2.34}
$$

and according to (2.10) we have

$$
|n(\boldsymbol{u}_1 - \boldsymbol{u}_2, \boldsymbol{u}_2, \boldsymbol{u}_1 - \boldsymbol{u}_2)| \leq C\|\boldsymbol{u}_2\|_1\|\boldsymbol{u}_1 - \boldsymbol{u}_2\|_1^2, \tag{2.35}
$$

and by (2.14) we can find μ such that

$$
|n(\boldsymbol{u}_1 - \boldsymbol{u}_2, \boldsymbol{g}_\mu, \boldsymbol{u}_1 - \boldsymbol{u}_2)| \leq \frac{\varepsilon_0}{4Re}\|\boldsymbol{u}_1 - \boldsymbol{u}_2\|_1^2. \tag{2.36}
$$

Now, we find upper bounds for $\boldsymbol{u}_1$ and $\boldsymbol{u}_2$. Testing the equation (2.22) with $\boldsymbol{u}$ results in

$$
\begin{aligned}
\frac{\varepsilon_0}{Re}\|\boldsymbol{u}\|_1^2 &\leq \|\boldsymbol{f}\|_{-1}\|\boldsymbol{u}\|_1 + \frac{\varepsilon_0}{Re}\|\boldsymbol{g}_\mu\|_1\|\boldsymbol{u}\|_1 + C\|\boldsymbol{g}_\mu\|_0\|\boldsymbol{u}\|_0 \\
&\quad + C\|\boldsymbol{g}_\mu\|_1^2\|\boldsymbol{u}\|_1 + C\|\beta\|_{0,\infty}\|\boldsymbol{g}_\mu\|_0\|\boldsymbol{u}\|_1^2 + C\|\beta\|_{0,\infty}\|\boldsymbol{g}_\mu\|_{0,4}^2\|\boldsymbol{u}\|_1 .
\end{aligned}
$$

From Sobolev embedding $H^1(\Omega) \hookrightarrow L^4(\Omega)$ we deduce for sufficiently small $\|\boldsymbol{g}_\mu\|_1$

$$
\|\boldsymbol{u}\|_1 \leq \frac{\|\boldsymbol{f}\|_{-1} + C_1\|\boldsymbol{g}_\mu\|_1 + C_2\|\boldsymbol{g}_\mu\|_1^2}{\dfrac{\varepsilon_0}{Re} - C_3\|\beta\|_{0,\infty}\|\boldsymbol{g}_\mu\|_1} =: C\big(\|\boldsymbol{g}_\mu\|_1, \|\boldsymbol{f}\|_{-1}\big). \tag{2.37}
$$

Putting (2.33)-(2.37) into (2.32) and using the inequality

$$\|\boldsymbol{f}\|_{-1}\|\boldsymbol{u}_1-\boldsymbol{u}_2\|_1 \le \frac{\varepsilon_0}{4Re}\|\boldsymbol{u}_1-\boldsymbol{u}_2\|_1^2+\frac{2Re}{\varepsilon_0}\|\boldsymbol{f}\|_{-1}^2$$

we obtain

$$\begin{aligned} 0 \ge & \frac{\varepsilon_0}{2Re}\|\boldsymbol{u}_1-\boldsymbol{u}_2\|_1^2-\frac{2Re}{\varepsilon_0}\|\boldsymbol{f}\|_{-1}^2-C\big(\|\boldsymbol{g}_\mu\|_1,\|\boldsymbol{f}\|_{-1}\big)\,\|\beta\|_{0,\infty}\|\boldsymbol{u}_1-\boldsymbol{u}_2\|_1^2 \\ & -\frac{\varepsilon_0}{4Re}\|\boldsymbol{u}_1-\boldsymbol{u}_2\|_1^2-C\big(\|\boldsymbol{g}_\mu\|_1,\|\boldsymbol{f}\|_{-1}\big)\|\boldsymbol{u}_1-\boldsymbol{u}_2\|_1^2 . \end{aligned} \tag{2.38}$$

For sufficiently small $\|\boldsymbol{g}_\mu\|_1$, $\|\boldsymbol{f}\|_{-1}$ the constant $C(\|\boldsymbol{g}_\mu\|_1,\|\boldsymbol{f}\|_{-1})$ in (2.37) gets small and consequently the right hand side of (2.38) is nonnegative. This implies $\boldsymbol{u}_1=\boldsymbol{u}_2$ and according to Theorem 2.9 is $p_1-p_2=0$. □

2.3 Finite element analysis

2.3.1 Discrete problem

For the finite element discretisation of (2.8), we are given a shape regular family $\{\mathcal{T}_h\}_{h>0}$ of decompositions of Ω into quadrilaterals ($n=2$) or hexahedrons ($n=3$). The diameter of the cell K will be denoted by h_K and the mesh size parameter h is defined by $h:=\max_{K\in\mathcal{T}_h} h_K$. Let $\boldsymbol{F}_K:\widehat{K}\to K$ be the multilinear reference mapping acting on the reference cell $\widehat{K}:=(-1,1)^n$. Now, we pay more attention to the strengthened shape regularity assumption given in [64]. Expanding $\boldsymbol{F}_K$ we get

$$\boldsymbol{F}_K(\hat{\boldsymbol{x}})=\boldsymbol{b}_K+\boldsymbol{B}_K\hat{\boldsymbol{x}}+\boldsymbol{G}_K(\hat{\boldsymbol{x}})$$

where

$$\boldsymbol{b}_K:=\boldsymbol{F}_K(\boldsymbol{0}),\ \boldsymbol{B}_K:=D\boldsymbol{F}_K(\boldsymbol{0}) \text{ and } \boldsymbol{G}_K(\hat{\boldsymbol{x}}):=\boldsymbol{F}_K(\hat{\boldsymbol{x}})-\boldsymbol{F}_K(\boldsymbol{0})-D\boldsymbol{F}_K(\boldsymbol{0})(\hat{\boldsymbol{x}}).$$

Let $\widehat{S}\subset\widehat{K}$ be the reference n-simplex having the following vertices $(0,\dots,0)$, $(1,0,\dots,0)$, ..., $(0,\dots,0,1)$. Its image via the affine mapping $\hat{x}\mapsto \boldsymbol{B}_K\hat{x}+\boldsymbol{b}_K$ is denoted by S_K. Furthermore, we denote by $h_{S_K}:=\operatorname{diam}(S_K)$ and ρ_{S_K} the diameter of S_K and the diameter of the largest ball inscribed into S_K, respectively. For each cell $K\in\mathcal{T}_h$ we define

$$\gamma_K:=\sup_{\hat{\boldsymbol{x}}\in\widehat{K}}\|\boldsymbol{B}_K^{-1}D\boldsymbol{F}_K(\hat{\boldsymbol{x}})-\boldsymbol{I}\| \tag{2.39}$$

which is a measure of the deviation of K from a parallelogram ($n=2$) or a parallelepiped ($n=3$) with respect to the matrix norm $\|\cdot\|$ induced by the Euclidian vector norm. We note that $\gamma_K = 0$ holds iff $\boldsymbol{F}_K$ is affine. We call a family of triangulations $\mathcal{T}_h$ shape regular if the conditions

$$\frac{h_{S_K}}{\rho_{S_K}} \leq C \tag{2.40}$$

and

$$\gamma_K \leq \gamma_0 < 1 \tag{2.41}$$

hold for all $K \in \mathcal{T}_h$. For this type of mesh the mapping $\boldsymbol{F}_K$ exhibits following properties

$$\|\boldsymbol{B}_K\| \leq C h_{S_K} \leq C h_K \,, \quad \|\boldsymbol{B}_K^{-1}\| \leq C h_{S_K}^{-1} \,, \quad \forall\, K \in \mathcal{T}_h \,, \tag{2.42}$$

$$\sup_{\hat{\boldsymbol{x}} \in \widehat{K}} \|D\boldsymbol{F}_K(\hat{\boldsymbol{x}})\| \leq (1+\gamma_K)\|\boldsymbol{B}_K\| \,, \quad \sup_{\boldsymbol{x} \in K} \|D\boldsymbol{F}_K^{-1}(\boldsymbol{x})\| \leq (1-\gamma_K)^{-1}\|\boldsymbol{B}_K^{-1}\| \quad \forall\, K \in \mathcal{T}_h \tag{2.43}$$

and

$$C h_K^n \leq n!(1-\gamma_K)^n \,|S_K| \leq |\det(D\boldsymbol{F}_K(\hat{\boldsymbol{x}}))| \leq n!(1+\gamma_K)^n \,|S_K| \leq C' h_K^n \quad \forall\, \hat{\boldsymbol{x}} \in \widehat{K} \,. \tag{2.44}$$

Moreover, the reference mapping $\boldsymbol{F}_K$ is bijective. We note that the above condition is more restrictive than the usual one for simplices

$$h_K/\rho_K < C \quad \forall\, K \in \mathcal{T}_h$$

where ρ_K denotes the diameter of the largest ball inscribed into K. We mention also that there are further variants of shape regularity condition which are frequently used in the literature, see the review articles [66, 65] for quadrilateral meshes.

Now, we want to prepare definitions of finite element spaces which will be used in the following. We denote discrete velocity and pressure finite element spaces by $\boldsymbol{X}_h$ and Q_h, respectively. Furthermore we define

$$\boldsymbol{X}_{h0} := \boldsymbol{X}_h \cap \boldsymbol{H}_0^1(\Omega) \,, \quad M_h = Q_h \cap L_0^2(\Omega) \,,$$

and

$$\boldsymbol{V}_h := \boldsymbol{X}_h \times M_h \,, \quad \boldsymbol{V}_{h0} := \boldsymbol{X}_{h0} \times M_h \,.$$

In the following we consider conforming finite element spaces, i.e. $\boldsymbol{V}_h \subset \boldsymbol{X} \times M$. We set

$$\begin{aligned}\tilde{A}(\boldsymbol{w}_h; \boldsymbol{u}_h, \boldsymbol{v}_h) := a(\boldsymbol{u}_h, \boldsymbol{v}_h) + c(\boldsymbol{u}_h, \boldsymbol{v}_h) + \tilde{n}(\boldsymbol{w}_h, \boldsymbol{u}_h, \boldsymbol{v}_h) \\ + d(\boldsymbol{w}_h; \boldsymbol{u}_h, \boldsymbol{v}_h) \,,\end{aligned}$$

where

$$\tilde{n}(\boldsymbol{w}_h, \boldsymbol{u}_h, \boldsymbol{v}_h) := \frac{1}{2}\big[n(\boldsymbol{w}_h, \boldsymbol{u}_h, \boldsymbol{v}_h) - n(\boldsymbol{w}_h, \boldsymbol{v}_h, \boldsymbol{u}_h)\big] \,.$$

We remark that according to Lemma 2.3 we have

$$\tilde{n}(\boldsymbol{w}, \boldsymbol{u}, \boldsymbol{v}) = n(\boldsymbol{w}, \boldsymbol{u}, \boldsymbol{v}) \qquad \text{if} \qquad \operatorname{div}(\varepsilon \boldsymbol{w}) = 0 \quad \text{and} \quad \boldsymbol{v} \in \boldsymbol{X}_0 \,. \tag{2.45}$$

The finite element discretisation of equation (2.8) leads to the following nonlinear problem

$$\begin{aligned}&\text{Find } (\boldsymbol{u}_h, p_h) \in \boldsymbol{V}_h \text{ with } \boldsymbol{u}_h|_\Gamma = \boldsymbol{g}_h|_\Gamma \quad \text{such that}\\ &\quad \tilde{A}(\boldsymbol{u}_h; \boldsymbol{u}_h, \boldsymbol{v}_h) - b(\boldsymbol{v}_h, p_h) + b(\boldsymbol{u}_h, q_h) = (\boldsymbol{f}, \boldsymbol{v}_h) \qquad \forall\, (\boldsymbol{v}_h, q_h) \in \boldsymbol{V}_{h0}\,. \end{aligned} \tag{2.46}$$

Here $\boldsymbol{g}_h \in \boldsymbol{X}_h$ is some approximation of extension of $\boldsymbol{g}$, e.g. $\boldsymbol{g}_h = \boldsymbol{i}_h \boldsymbol{g}$ with $\boldsymbol{i}_h \boldsymbol{g}$ being an appropriate finite element interpolant of $\boldsymbol{g}$.
First, we pay attention to the fixed point linearisation

$$\begin{aligned}&\text{For given } \boldsymbol{u}_h^{old} \in \boldsymbol{X}_h \text{ find } (\boldsymbol{u}_h, p_h) \in \boldsymbol{V}_h \text{ with } \boldsymbol{u}_h|_\Gamma = \boldsymbol{g}_h|_\Gamma \quad \text{such that}\\ &\quad \tilde{A}(\boldsymbol{u}_h^{old}; \boldsymbol{u}_h, \boldsymbol{v}_h) - b(\boldsymbol{v}_h, p_h) + b(\boldsymbol{u}_h, q_h) = (\boldsymbol{f}, \boldsymbol{v}_h) \qquad \forall\, (\boldsymbol{v}_h, q_h) \in \boldsymbol{V}_{h0}\,. \end{aligned} \tag{2.47}$$

We want to study its solvability in the spirit of mixed finite elements, see [30]. To this end, we define the space

$$\boldsymbol{W}_h := \{\boldsymbol{w}_h \in \boldsymbol{X}_{h0} : \quad b(\boldsymbol{w}_h, q_h) = 0 \quad \forall\, q_h \in Q_h\}$$

of finite element functions satisfying the modified divergence constraint in the discrete sense. Note that in general $\boldsymbol{W}_h \not\subset \boldsymbol{W}$. The linear saddle point problem (2.47) formulated on the subspace $\boldsymbol{W}_h \subset \boldsymbol{X}_{h0}$ simplifies to an elliptic one. Then, the existence and uniqueness of the discrete velocity follow obviously from the Lax-Milgram Lemma.

2.3.2 Verifying the discrete inf-sup condition

It is well known in mixed finite element method for incompressible fluids that velocity and pressure have to be approximated by suitable finite element pairs which satisfy the Babuška-Brezzi compatibility condition, otherwise the pressure approximation can be deteriorated by unphysical oscillations. One can expect similar difficulties when solving the discrete saddle point problem (2.47). The general result concerning this issue is established by the following lemma.

Lemma 2.11 *Let the pair $(\boldsymbol{X}_{h0}, M_h)$ satisfy the discrete inf-sup condition*

$$\exists\, \gamma > 0 : \qquad \inf_{q_h \in M_h} \sup_{\boldsymbol{v}_h \in \boldsymbol{X}_{h0}} \frac{b(\boldsymbol{v}_h, q_h)}{|\boldsymbol{v}_h|_1 \, \|q_h\|_0} \geq \gamma \qquad \forall\, h > 0\,. \tag{2.48}$$

Then the discrete problem (2.47) has a unique solution $(\boldsymbol{u}_h, p_h) \in \boldsymbol{X}_h \times M_h$.

Proof. See the proof of [30, Theorem 1.1, §1, Chapter II]. □

Remark 2.12 *The inf-sup condition (2.48) can be reformulated as follows*

$$\exists\, \gamma > 0 \quad \forall\, q_h \in M_h \quad \exists\, \boldsymbol{v}_h \in \boldsymbol{X}_{h0} : \qquad\qquad b(\boldsymbol{v}_h, q_h) = \|q_h\|_0^2 \,, \qquad |\boldsymbol{v}_h|_1 \le \frac{1}{\gamma}\|q_h\|_0 \qquad \forall\, h > 0\,. \tag{2.49}$$

Remark 2.13 *In the case of $\gamma = \gamma(h) > 0$ the existence of discrete pressure can be also guaranteed. However, the optimal error bounds can not be derived using such type of inf-sup constant.*

Let $\widehat{P}_k$ denote the space of polynomials of degree at most $k \ge 0$, and $\widehat{Q}_k$ be the space of polynomials of degree at most $k \ge 0$ in each variable separately. In the following we consider a family of mapped finite elements with discontinuous pressure. Let

$$P_k(K) := \{v = \hat{v} \circ \boldsymbol{F}_K^{-1} \, : \hat{v} \in \widehat{P}_k\} \quad \text{and} \quad Q_k(K) := \{v = \hat{v} \circ \boldsymbol{F}_K^{-1} \, : \hat{v} \in \widehat{Q}_k\}$$

be the local finite element spaces defined on a given cell K via reference mapping $\boldsymbol{F}_K$. The vector valued counterparts of them we denote by bold faced symbols $\boldsymbol{Q}_k(K) := [Q_k(K)]^n$. For $k \ge 2$ we look for the pressure and velocity approximations in the following finite element spaces

$$\boldsymbol{X}_h := \{\boldsymbol{v} \in C^0(\Omega) \, : \, \boldsymbol{v}|_K \in \boldsymbol{Q}_k(K) \quad \forall\, K \in \mathcal{T}_h\}$$

and

$$M_h := \{q \in L_0^2(\Omega) \, : \, q|_K \in P_{k-1}(K) \quad \forall\, K \in \mathcal{T}_h\}\,,$$

respectively. Note that the inf-sup condition for this finite element pair in case of the Stokes–Problem in two and three dimensions has been proved in [64] by means of the macro-element technique of Boland and Nicolaides [10]. We extend these results in order to prove the inf-sup condition for the bilinear form corresponding to the modified divergence constraint of model equations (2.5). Let the domain Ω be decomposed into non-overlapping, open domains Ω_r, $r = 1, \ldots, R$ with Lipschitz continuous boundary,

$$\overline{\Omega} = \bigcup_{r=1}^{R} \overline{\Omega}_r \,, \qquad \Omega_r \cap \Omega_s = \emptyset \,, \qquad r \ne s\,.$$

Furthermore we define the following finite element subspaces on the subdomains Ω_r

$$\begin{aligned} \boldsymbol{X}_h(\Omega_r) &: = \{v|_{\Omega_r} \, : \, \boldsymbol{v} \in \boldsymbol{X}_{h0}\,, \ \boldsymbol{v} = 0 \text{ in } \Omega \setminus \Omega_r\} \\ Q_h(\Omega_r) &: = \{q|_{\Omega_r} \, : \, q \in Q_h\} \\ M_h(\Omega_r) &: = Q_h \cap L_0^2(\Omega_r) \end{aligned}$$

and

$$\overline{M}_h := \{q \in L_0^2(\Omega) \, : \, q|_{\Omega_r} = const\,, \ r = 1, \ldots, R\}\,.$$

Let us formulate the local inf-sup condition

$$\exists\,\lambda>0\ \ \forall\,r=1,\dots,R\,,$$
$$\forall\,q_h\in M_h(\Omega_r):\qquad \sup_{\boldsymbol{v}_h\in\boldsymbol{X}_h(\Omega_r)}\frac{b(\boldsymbol{v}_h,q_h)}{|\boldsymbol{v}_h|_{1,\Omega_r}}\ge\lambda\|q_h\|_{0,\Omega_r}\,. \tag{2.50}$$

The following lemma gives a connection between the local and global inf-sup conditions in order to establish stable finite element pairs.

Lemma 2.14 *Assume that the bilinear form $b:\ \boldsymbol{X}_0\times M\to\mathbb{R}$ is continuous and satisfies $b(q_h,\boldsymbol{v}_h)=0$ for all piecewise constant pressures q_h and $\boldsymbol{v}_h\in\boldsymbol{X}_{h0}$ with $\boldsymbol{v}_h|_{\Omega_r}\in\boldsymbol{X}_h(\Omega_r)$, $r=1,\dots,R$. Let the local inf-sup condition (2.50) be satisfied with a constant $\lambda>0$ independent of r and the mesh size parameter h. Furthermore, we assume that there exists a subspace $\overline{\boldsymbol{X}}_h\subset\boldsymbol{X}_{h0}$ such that $(\overline{\boldsymbol{X}}_h,\overline{M}_h)$ is globally inf-sup stable with a constant $\bar\gamma>0$ independent of h. Then, there exists a constant $\gamma>0$ independent of h such that the pair $(\boldsymbol{X}_{h0},M_h)$ satisfies the global inf-sup condition (2.48).*

Proof. We follow the proof of [30, Theorem 1.12, §1, Chapter II] and adapt it to the abstract continuous bilinear form $b(\cdot,\cdot)$. From the local orthogonal decomposition $Q_h(\Omega_r)=M_h(\Omega_r)\oplus\mathbb{R}$ we deduce that each function $q_h\in M_h$ can be split Ω_r as follows

$$q_h=\tilde q_h+\bar q_h$$

where $\bar q_h|_{\Omega_r}:=\dfrac{(q_h,1)_{\Omega_r}}{|\Omega_r|}$ and $\tilde q_h|_{\Omega_r}\in M_h(\Omega_r)$. From the orthogonal decomposition follows obviously the global relation

$$\|q_h\|_0^2=\|\tilde q_h\|_0^2+\|\bar q_h\|_0^2\,. \tag{2.51}$$

Now, we reformulate the local inf-sup condition (2.50) by analogy to Remark 2.12 and state that there exists a function $\tilde{\boldsymbol{v}}_r\in\boldsymbol{X}_h(\Omega_r)$ such that

$$b(\tilde{\boldsymbol{v}}_r,\tilde q_r)=\|\tilde q_r\|_{0,\Omega_r}^2\qquad\text{and}\qquad|\tilde{\boldsymbol{v}}_r|_{1,\Omega_r}\le\frac{1}{\lambda}\|\tilde q_r\|_{0,\Omega_r} \tag{2.52}$$

where $\tilde q_r=\tilde q_h|_{\Omega_r}$. Applying the same argument to the globally inf-sup stable pair $(\overline{\boldsymbol{X}}_h,\overline{M}_h)$ we conclude that there exists a function $\bar{\boldsymbol{v}}_h\in\overline{\boldsymbol{X}}_h$ such that

$$b(\bar{\boldsymbol{v}}_h,\bar q_h)=\|\bar q_h\|_0^2\qquad\text{and}\qquad|\bar{\boldsymbol{v}}_h|_1\le\frac{1}{\bar\gamma}\|\bar q_h\|_0\,. \tag{2.53}$$

Let us define for $t>0$ the function $\boldsymbol{v}_h\in\boldsymbol{X}_{h0}$ defined by

$$\boldsymbol{v}_h:=\tilde{\boldsymbol{v}}_h+t\bar{\boldsymbol{v}}_h$$

where $t>0$. We adjust the parameter $t>0$ such that the pair $(\boldsymbol{v}_h,q_h)$ obeys the global inf-sup condition reformulated in Remark 2.12. To this end, we evaluate

$$b(\boldsymbol{v}_h,q_h)=b(\tilde{\boldsymbol{v}}_h,\tilde q_h)+b(\tilde{\boldsymbol{v}}_h,\bar q_h)+tb(\bar{\boldsymbol{v}}_h,\tilde q_h)+tb(\bar{\boldsymbol{v}}_h,\bar q_h)\,. \tag{2.54}$$

We have

$$b(\tilde{\boldsymbol{v}}_h, \bar{q}_h) = 0 \tag{2.55}$$

due to the assumptions on $b(\cdot,\cdot)$. From (2.52) and (2.53) follows

$$b(\tilde{\boldsymbol{v}}_h, \tilde{q}_h) = \|\tilde{q}_h\|_0^2 \tag{2.56}$$

and

$$b(\bar{\boldsymbol{v}}_h, \bar{q}_h) = \|\bar{q}_h\|_0^2 \,. \tag{2.57}$$

The continuity of the bilinear form $b(\cdot,\cdot)$, i.e.,

$$|b(\boldsymbol{v}, q)| \leq M_b |\boldsymbol{v}|_1 \|q\|_0 \qquad \forall\, \boldsymbol{v} \in \boldsymbol{X}_0 \quad \forall\, q \in M \,,$$

and (2.53) implies

$$b(\bar{\boldsymbol{v}}_h, \tilde{q}_h) \leq \frac{M_b}{\bar{\gamma}} \|\tilde{q}_h\|_0 \, \|\bar{q}_h\|_0 \,. \tag{2.58}$$

Collecting (2.54)-(2.58) and employing the inequality

$$\|\tilde{q}_h\|_0 \|\bar{q}_h\|_0 \leq \vartheta \|\tilde{q}_h\|_0^2 + \frac{1}{4\vartheta} \|\bar{q}_h\|_0^2 \qquad \forall\, \vartheta > 0 \,,$$

we get for all $\vartheta > 0$ the lower bound

$$b(\boldsymbol{v}_h, q_h) \geq \left\{ 1 - \frac{M_b t \vartheta}{\bar{\gamma}} \right\} \|\tilde{q}_h\|_0^2 + t \left\{ 1 - \frac{M_b}{4\vartheta\bar{\gamma}} \right\} \|\bar{q}_h\|_0^2 \,. \tag{2.59}$$

Choosing $\vartheta = \frac{\bar{\gamma}}{2t\sqrt{M_b}}$ and $t = \frac{\bar{\gamma}}{M_b}$ in the above estimate, we obtain

$$b(\boldsymbol{v}_h, q_h) \geq \min\left\{ \frac{1}{2}, \frac{\bar{\gamma}^2}{2M_b} \right\} \|q_h\|_0^2 \,. \tag{2.60}$$

Furthermore, the following bound

$$|\boldsymbol{v}_h|_1 \leq |\tilde{\boldsymbol{v}}_h|_1 + t|\bar{\boldsymbol{v}}_h|_1 \leq \frac{1}{\lambda} \|\tilde{q}_h\|_0 + \frac{\bar{\gamma}}{M_b} \|\bar{q}_h\|_0 \leq \left\{ \left(\frac{1}{\lambda}\right)^2 + \left(\frac{\bar{\gamma}}{M_b}\right)^2 \right\}^{1/2} \|q_h\|_0 \,. \tag{2.61}$$

holds according to (2.51)-(2.53). From (2.60) and (2.61) we conclude

$$\frac{b(\boldsymbol{v}_h, q_h)}{|\boldsymbol{v}_h|_1} \geq \gamma \|q_h\|_0 \qquad \forall\, h > 0$$

with the inf-sup constant

$$\gamma = \frac{\min\left\{ \frac{1}{2}, \frac{\bar{\gamma}^2}{2M_b} \right\}}{\left\{ \left(\frac{1}{\lambda}\right)^2 + \left(\frac{\bar{\gamma}}{M_b}\right)^2 \right\}^{1/2}} \,. \tag{2.62}$$

□

Remark 2.15 *We observe that the bilinear form* $b(\boldsymbol{v},\boldsymbol{q}) = \big(div(\varepsilon\boldsymbol{v}), q\big)$ *satisfies assumptions of Lemma 2.14. Indeed,* $b(\cdot,\cdot)$ *is continuous with the constant* $M_b = C\{\varepsilon_1 + |\varepsilon|_{1,3}\}\sqrt{n}$ *since*

$$|b(\boldsymbol{v},q)| \leq \sqrt{n}\,|\varepsilon\boldsymbol{v}|_1\,\|q\|_0 \leq C\{\varepsilon_1 + |\varepsilon|_{1,3}\}\sqrt{n}\,|\boldsymbol{v}|_1\|q\|_0$$

due to Theorem 2.2 and Poincaré inequality. It holds also $b(\boldsymbol{v}_h, q_h) = 0$ *for all piecewise constant pressures* q_h *and* $\boldsymbol{v}_h \in \boldsymbol{X}_{h0}$ *with* $\boldsymbol{v}_h|_{\Omega_r} \in \boldsymbol{X}_h(\Omega_r)$*,* $r = 1,\ldots,R$*, since by virtue of Gauss theorem we have*

$$b(\boldsymbol{v}_h, q_h) = \big(div(\varepsilon\boldsymbol{v}_h), q_h\big) = q_h|_{\Omega_r}\int\limits_{\Omega_r} div(\varepsilon\boldsymbol{v}_h)\,dx$$

$$= q_h|_{\Omega_r}\int\limits_{\partial\Omega_r} \varepsilon\boldsymbol{v}_h\cdot\boldsymbol{n}ds = 0$$

whereby $\boldsymbol{n}$ *denotes the outer normal vector on* Ω_r*.*

Now, we want to find a pair of subspaces $(\overline{\boldsymbol{X}}_h, \overline{M}_h)$ which satisfies the global inf-sup condition. We recall the notation given in [64] in order to establish appropriate enrichment of Q_1-velocity which together with the piecewise constant pressure results in an inf-sup stable pair. Let $\mathcal{E}(K)$ be a set of all $(n-1)$-dimensional faces of an element $K \in \mathcal{T}$ and $\boldsymbol{n}_K^E$ the unit normal vector to the face $E \in \mathcal{E}(K)$. Furthermore, we denote by $\mathcal{E}$ the set of all faces $E \in \mathcal{E}(K)$ of all elements $K \in \mathcal{T}$, by $\mathcal{E}(\Gamma)$ we denote the set of all faces located at the domain boundary Γ. The reference cell $\widehat{E} = \boldsymbol{F}_K^{-1}(E)$ corresponds to the face $E \in \mathcal{E}(K)$. We denote by $\boldsymbol{n}_E^0 = \boldsymbol{n}_E(\boldsymbol{m}_E)$ the unit normal vector $\boldsymbol{n}_E$ at the point $\boldsymbol{m}_E$ being the image of the barycentre of the reference face $\widehat{E}$ under the reference transformation $\boldsymbol{F}_K$, i.e. $\boldsymbol{m}_E = \boldsymbol{F}_K(\hat{\boldsymbol{m}}_{\hat{E}})$. Note that the above definition is justified by the fact that the unit normal vector in three dimensions is in general no longer constant. Let us introduce the scalar function $\psi_E \in Q_2(K)$ which is uniquely defined on $K = K(E)$ (adjacent cell to the face E) by nine nodes $\boldsymbol{a}_j \in K$

$$\psi_E(\boldsymbol{a}_j) := \begin{cases} 1 & \text{if} \quad \boldsymbol{a}_j = \boldsymbol{m}_E\,, \\ 0 & \text{otherwise}\,. \end{cases}$$

We define the face bubble function by

$$\boldsymbol{\Phi}_E(\boldsymbol{x}) := \psi_E(\boldsymbol{x})\boldsymbol{n}_E^0 \quad \forall\,\boldsymbol{x} \in \Omega\,.$$

Within the extended local finite element space

$$\boldsymbol{Q}_1^+(K) = \boldsymbol{Q}_1(K) + \mathrm{span}\{\boldsymbol{\Phi}_E\,,\quad E \in \mathcal{E}(K)\}$$

we define the subspace

$$\overline{\boldsymbol{X}}_h = \{v \in \boldsymbol{C}^0(\Omega) : \quad \boldsymbol{v}|_K \in \boldsymbol{Q}_1^+(K) \quad \forall\, K \in \mathcal{T}_h\ \boldsymbol{v}|_{\partial\Omega} = 0\}\,. \tag{2.63}$$

The proof of global inf-sup stability of the pair $(\overline{\boldsymbol{X}}_h, \overline{M}_h)$ involves the following lemma.

Lemma 2.16 *The global inf-sup condition (2.48) holds if and only if there exists an operator $\boldsymbol{\Pi}_h \in \mathcal{L}(\boldsymbol{X}_0, \boldsymbol{X}_{h0})$ satisfying:*

$$b(\boldsymbol{v}, q_h) = b(\boldsymbol{\Pi}_h \boldsymbol{v}, q_h) \quad \forall\, q_h \in M_h \quad \forall\, \boldsymbol{v} \in \boldsymbol{X}_0\,, \tag{2.64}$$

$$|\boldsymbol{\Pi}_h \boldsymbol{v}|_1 \le C |\boldsymbol{v}|_1 \quad \forall\, \boldsymbol{v} \in \boldsymbol{X}_0 \tag{2.65}$$

with a constant $C > 0$ independent of $h > 0$.

Proof. See the proof [30, Lemma 1.1, §1, Chapter II] for an abstract bilinear form $b(\cdot,\cdot)$.
□

Remark 2.17 *From the proof of [30, Lemma 1.1, §1, Chapter II] follows that the discrete inf-sup constant in condition (2.48) is determined by*

$$\gamma = \gamma_\varepsilon / C$$

where $C > 0$ is the stability constant in (2.65) and $\gamma_\varepsilon > 0$ is the inf-sup constant in condition (2.31) for the continuous spaces.

Now, we construct the H^1 stable operator $\boldsymbol{\Pi}_h$ explicitly. To this end, let $\boldsymbol{i}_h : \boldsymbol{H}^1(\Omega) \to \boldsymbol{X}_h$ be the interpolation operator of Scott–Zhang type. This type of interpolation operators defined for non-smooth functions on affine equivalent meshes and preserving Dirichlet homogeneous boundary condition was introduced in [70], and extended in [35] to the case of shape regular meshes including hanging nodes. A good source which supplies construction principles of Scott–Zhang operator for higher order elements is the text book of Ern and Guermond [24]. Following ideas therein we explain briefly the construction of Scott–Zhang operator $\boldsymbol{i}_h : \boldsymbol{H}^1(\Omega) \to \boldsymbol{X}_h$ for vector valued functions. Let $\{\boldsymbol{a}_1, \dots, \boldsymbol{a}_N\}$ be the Lagrange nodal points and let $\{\varphi_1, \dots, \varphi_N\}$ be the set of the global shape functions of $\boldsymbol{X}_h$. With each node $\boldsymbol{a}_i$ we associate

$$\sigma_i := \begin{cases} K_i\,, & \text{if } \boldsymbol{a}_i \text{ is in interior of } K_i \in \mathcal{T}_h\,, \\ E_i\,, & \text{if } \boldsymbol{a}_i \notin \partial\Omega \text{ and } \boldsymbol{a}_i \text{ is on the face } E_i \in \mathcal{E}_h\,, \\ E_i \subset \partial\Omega\,, & \text{if } \boldsymbol{a}_i \in \partial\Omega \text{ and } \boldsymbol{a}_i \text{ is on the face } E_i \in \mathcal{E}_h\,. \end{cases}$$

We note that the choice of E_i in the definition of σ_i is not unique. Let n_i be the number of nodes belonging to σ_i. We denote the local shape functions restricted to σ_i and associated with the nodes lying in σ_i by $\{\varphi_{i,1}, \dots, \varphi_{i,n_i}\}$, and set conventionally $\varphi_{i,1} = \varphi_i$, $i = 1, \dots, N$. For each such set of nodal functions on σ_i we construct a dual $L^2(\sigma_i)$ basis $\{\psi_{i,1}, \dots, \psi_{1,n_i}\}$ with

$$\int_{\sigma_i} \varphi_{i,q} \psi_{i,r} = \delta_{qr}\,, \qquad 1 \le q, r \le n_i\,.$$

Furthermore, we define nodal functionals $N_i : \boldsymbol{H}^1(\Omega) \to \mathbb{R}$ by

$$N_i[\boldsymbol{v}] := \int\limits_{\sigma_i} \boldsymbol{v}\boldsymbol{\psi}_i \,.$$

Then, the Scott-Zhang interpolation operator is defined for $\boldsymbol{v} \in \boldsymbol{H}^1(\Omega)$ by the condition

$$N_i[\boldsymbol{v} - \boldsymbol{i}_h\boldsymbol{v}] = 0 \quad \forall i = 1, \ldots, N$$

which implies the following representation

$$(\boldsymbol{i}_h\boldsymbol{v})(\boldsymbol{x}) = \sum_{i=1}^{N} N_i[\boldsymbol{v}]\, \boldsymbol{\varphi}_i(\boldsymbol{x})$$

Obviously, the Scott–Zhang operator preserves homogeneous boundary conditions, i.e., $\boldsymbol{v}|_{\partial\Omega} = \boldsymbol{0} \;\Rightarrow\; (\boldsymbol{i}_h\boldsymbol{v})\,|_{\partial\Omega} = \boldsymbol{0}$, and furthermore satisfies $\boldsymbol{i}_h\boldsymbol{v}_h = \boldsymbol{v}_h$ for all $\boldsymbol{v}_h \in \boldsymbol{X}_h$. It has been proven that this kind of interpolation operator is H^1 stable

$$|\boldsymbol{i}_h\boldsymbol{v}|_{1,K} \le C|\boldsymbol{v}|_{1,\omega(K)} \quad \forall\, \boldsymbol{v} \in \boldsymbol{H}^1(\Omega) \tag{2.66}$$

and satisfies on each $K \in \mathcal{T}_h$ the approximation property

$$\|\boldsymbol{v} - \boldsymbol{i}_h\boldsymbol{v}\|_{0,K} + h_K|\boldsymbol{v} - \boldsymbol{i}_h\boldsymbol{v}|_{1,K} \le Ch_K^l|\boldsymbol{v}|_{l,\omega(K)} \quad \forall\, \boldsymbol{v} \in \boldsymbol{H}^l(\omega(K)) \quad 1 \le l \le k+1\,, \tag{2.67}$$

where $\omega(K)$ denotes a certain local neighbourhood of K. Furthermore we define the global operator $\boldsymbol{I}_h : \boldsymbol{X}_0 \to \overline{\boldsymbol{X}}_h$ by

$$\boldsymbol{I}_h\boldsymbol{v} = \sum_{E\in\mathcal{E}} \frac{\langle \boldsymbol{v}\cdot\boldsymbol{n}_E, \varepsilon\rangle_E}{\langle \boldsymbol{\Phi}_E\cdot\boldsymbol{n}_E, \varepsilon\rangle_E}\boldsymbol{\Phi}_E \,.$$

Integrating by parts for $q_h \in \overline{M}_h$, $q_K := q_h|_K$ and taking into account the fact that the bubble function $\boldsymbol{\Phi}_E$ vanishes on $\mathcal{E}(K)\setminus E$, implies

$$\begin{aligned}
(\operatorname{div}(\varepsilon\boldsymbol{I}_h\boldsymbol{v}), q_h) &= \sum_{K\in\mathcal{T}_h} \big(\operatorname{div}(\varepsilon\boldsymbol{I}_h\boldsymbol{v}), q_K\big)_K \\
&= \sum_{K\in\mathcal{T}_h}\sum_{E\in\mathcal{E}(K)} q_K\langle \boldsymbol{I}_h\boldsymbol{v}\cdot\boldsymbol{n}_K^E, \varepsilon\rangle_E \\
&= \sum_{K\in\mathcal{T}_h}\sum_{E\in\mathcal{E}(K)}\sum_{E'\in\mathcal{E}} q_K\frac{\langle \boldsymbol{v}\cdot\boldsymbol{n}_{E'}, \varepsilon\rangle_{E'}}{\langle \boldsymbol{\Phi}_{E'}\cdot\boldsymbol{n}_{E'}, \varepsilon\rangle_{E'}}\langle \boldsymbol{\Phi}_{E'}\cdot\boldsymbol{n}_K^E, \varepsilon\rangle_E \\
&- \sum_{K\in\mathcal{T}_h}\sum_{E\in\mathcal{E}(K)} \langle \boldsymbol{v}\cdot\boldsymbol{n}_K^E, \varepsilon q_K\rangle_E = \sum_{K\in\mathcal{T}_h} \big(\operatorname{div}(\varepsilon\boldsymbol{v}), q_K\big)_K \\
&= \big(\operatorname{div}(\varepsilon\boldsymbol{v}), q_h\big)\,.
\end{aligned} \tag{2.68}$$

Employing the estimates from [36]

$$\langle \mathbf{\Phi}_E \cdot \boldsymbol{n}_E, \varepsilon \rangle_E \geq C\varepsilon_0 h_K^{n-1} \qquad \forall\, E \in \mathcal{E}(K)\,, \tag{2.69}$$

$$\|\boldsymbol{v}\|_{0,E} \leq C h_K^{1/2} \{ h_K^{-1} \|\boldsymbol{v}\|_{0,K} + |\boldsymbol{v}|_{1,K} \} \qquad \forall\, \boldsymbol{v} \in \boldsymbol{H}^1(K) \tag{2.70}$$

and

$$|\mathbf{\Phi}_E|_{1,K} \leq C h_K^{n/2-1} \qquad \forall\, K \in \mathcal{T}_h \tag{2.71}$$

yields for shape regular meshes

$$|\boldsymbol{I}_h \boldsymbol{v}|_{1,\Omega}^2 \leq C \sum_{K \in \mathcal{T}_h} \{ h_K^{-2} \|\boldsymbol{v}\|_{0,K}^2 + |\boldsymbol{v}|_{1,K}^2 \} \qquad \forall\, \boldsymbol{v} \in \boldsymbol{H}_0^1(\Omega)\,. \tag{2.72}$$

Now, we define the interpolation operator $\mathbf{\Pi}_h : \boldsymbol{H}_0^1(\Omega) \to \boldsymbol{X}_{h0}$ by

$$\mathbf{\Pi}_h \boldsymbol{v} := \boldsymbol{i}_h \boldsymbol{v} + \boldsymbol{I}_h(\boldsymbol{v} - \boldsymbol{i}_h \boldsymbol{v}) \qquad \forall\, \boldsymbol{v} \in \boldsymbol{H}_0^1(\Omega)\,. \tag{2.73}$$

Its main properties are summarised in the following lemma.

Lemma 2.18 *For the shape regular meshes the operator $\mathbf{\Pi}_h$ defined in (2.73) is H^1 stable in the sense of (2.65) and satisfies the condition (2.64).*

Proof. From the stability of $\boldsymbol{i}_h$ in (2.66) and of $\boldsymbol{I}_h$ in (2.72) we obtain for the operator $\mathbf{\Pi}_h$ defined in (2.73) the following estimate

$$\begin{aligned} |\mathbf{\Pi}_h \boldsymbol{v}|_{1,\Omega}^2 &\leq 2|\boldsymbol{i}_h \boldsymbol{v}|_{1,\Omega}^2 + 2|\boldsymbol{I}_h(\boldsymbol{v} - \boldsymbol{i}_h \boldsymbol{v})|_{1,\Omega}^2 \\ &\leq C \Big\{ |\boldsymbol{v}|_{1,\Omega}^2 + \sum_{K \in \mathcal{T}_h} \{ h_K^{-2} \|\boldsymbol{v} - \boldsymbol{i}_h \boldsymbol{v}\|_{0,K}^2 + |\boldsymbol{v} - \boldsymbol{i}_h \boldsymbol{v}|_{1,K}^2 \} \Big\} \\ &\leq C |\boldsymbol{v}|_{1,\Omega}^2 \,. \end{aligned} \tag{2.74}$$

In (2.68) we have already shown that the operator $\mathbf{\Pi}_h$ satisfies the condition (2.64). □

The main result concerning the global inf-sup stability of the pair $(\overline{\boldsymbol{X}}_h, \overline{M}_h)$ is established by the following lemma.

Lemma 2.19 *The global inf-sup condition (2.48) holds for the finite element pair $(\overline{\boldsymbol{X}}_h, \overline{M}_h)$ on shape regular meshes.*

Proof. Apply Lemma 2.16 and 2.18. □

Now, we are able to state our main result.

Theorem 2.20 *Let $\mathcal{T}_h$ be shape regular. Then, the pair $(\boldsymbol{X}_h, M_h)$ of mapped (Q_k, P_{k-1}^{disc}) finite element spaces satisfies the discrete inf-sup condition (2.48).*

Proof. Let the partition of Ω be $\mathcal{T}_h$ itself. Therefore each subdomain Ω_r is a certain cell $K \in \mathcal{T}_h$. From the previous lemma follows that the pair $(\overline{\boldsymbol{X}}_h, \overline{M}_h)$ satisfies the global inf-sup condition (2.48). In view of Lemma 2.14 it remains to prove the local inf-sup condition (2.50). To this end, we define for the arbitrary pressure $q \in M_h(K)$, $q \neq 0$, the velocity $\hat{\boldsymbol{v}} : \widehat{K} \to \mathbb{R}$ by

$$\hat{\boldsymbol{v}}(\hat{\boldsymbol{x}}) := -\boldsymbol{B}_K^{-T} \cdot \big(\hat{\nabla}(q \circ \boldsymbol{F}_K)\big)(\hat{\boldsymbol{x}}) \cdot \hat{b}(\hat{\boldsymbol{x}}) \tag{2.75}$$

where $\boldsymbol{F}_K$ is the multilinear reference transformation explained in Subsection 2.3.1, $\hat{\nabla} = \left(\frac{\partial}{\partial \hat{x}_1}, \ldots, \frac{\partial}{\partial \hat{x}_n}\right)^T$ denotes the column nabla operator with respect to the reference coordinates $\hat{x}_1, \ldots, \hat{x}_n$, and

$$\hat{b}(\hat{\boldsymbol{x}}) := \prod_{i=1}^{n} \left(1 - \hat{x}_i^2\right) \tag{2.76}$$

stands for the bubble function which is positive in the interior of $\widehat{K}$. Obviously, we have $\hat{\boldsymbol{v}}|_{\partial \widehat{K}} = \boldsymbol{0}$, $\hat{\boldsymbol{v}} \in \widehat{\boldsymbol{Q}}_k$ due to $\hat{\nabla}(q \circ \boldsymbol{F}_K) \in \widehat{\boldsymbol{P}}_{k-2}$, and therefore $\boldsymbol{v} = \hat{\boldsymbol{v}} \circ \boldsymbol{F}_K^{-1} \in \boldsymbol{X}_h(K)$. By virtue of the chain rule we have

$$\big(\hat{\nabla}\hat{q}\big)(\hat{\boldsymbol{x}}) = \boldsymbol{B}_K^T \cdot \left\{\boldsymbol{I} + \boldsymbol{B}_K^{-1} D\boldsymbol{G}_K(\hat{\boldsymbol{x}})\right\}^T \cdot \big(\nabla q\big)\big(\boldsymbol{F}_K(\hat{\boldsymbol{x}})\big) \tag{2.77}$$

where $\hat{q} = q \circ \boldsymbol{F}_K$. Then, integrating by parts, using reference transformation and employing (2.75) yields

$$\begin{aligned}
&\big(\operatorname{div}(\varepsilon \boldsymbol{v}), q\big)_K = -(\varepsilon \boldsymbol{v}, \nabla q)_K \\
&= -\int_{\widehat{K}} \varepsilon\big(\boldsymbol{F}_K(\hat{\boldsymbol{x}})\big) \cdot \hat{\boldsymbol{v}}^T(\hat{\boldsymbol{x}}) \cdot \big(\nabla q\big)\big(\boldsymbol{F}_K(\hat{\boldsymbol{x}})\big) \cdot |\det D\boldsymbol{F}_K(\hat{\boldsymbol{x}})| \, d\hat{\boldsymbol{x}} \\
&= \int_{\widehat{K}} \varepsilon\big(\boldsymbol{F}_K(\hat{\boldsymbol{x}})\big) \cdot \hat{b}(\hat{\boldsymbol{x}}) \cdot \big(\nabla q\big)^T\big(\boldsymbol{F}_K(\hat{\boldsymbol{x}})\big) \cdot \left\{\boldsymbol{I} + \boldsymbol{B}_K^{-1} D\boldsymbol{G}_K(\hat{\boldsymbol{x}})\right\} \cdot \big(\nabla q\big)\big(\boldsymbol{F}_K(\hat{\boldsymbol{x}})\big) \cdot |\det D\boldsymbol{F}_K(\hat{\boldsymbol{x}})| \, d\hat{\boldsymbol{x}} .
\end{aligned} \tag{2.78}$$

Next, the shape regularity assumptions (2.40) and (2.41) imply

$$\boldsymbol{z}^T \left\{\boldsymbol{I} + \boldsymbol{B}_K^{-1} D\boldsymbol{G}_K(\hat{\boldsymbol{x}})\right\} \boldsymbol{z} \geq (1 - \gamma_0) \, \|\boldsymbol{z}\|^2 \qquad \forall\, \boldsymbol{z} \in \mathbb{R}^n .$$

Consequently, we obtain for $\hat{b} \geq 0$ and $\varepsilon \geq \varepsilon_0 > 0$ the lower bound of (2.78)

$$\big(\operatorname{div}(\varepsilon \boldsymbol{v}), q\big)_K \geq (1 - \gamma_0)\varepsilon_0 \int_{\widehat{K}} \hat{b}(\hat{\boldsymbol{x}}) \cdot \left|\big(\nabla q\big)\big(\boldsymbol{F}_K(\hat{\boldsymbol{x}})\big)\right|^2 \cdot |\det D\boldsymbol{F}_K(\hat{\boldsymbol{x}})| \, d\hat{\boldsymbol{x}} . \tag{2.79}$$

From (2.77), (2.40) and (2.41) we get for the shape regular meshes

$$\frac{1}{(1 + \gamma_0)^2 \|\boldsymbol{B}_K\|^2} \left|(\hat{\nabla}\hat{q})(\hat{\boldsymbol{x}})\right|^2 \leq \left|\big(\nabla q\big)\big(\boldsymbol{F}_K(\hat{\boldsymbol{x}})\big)\right|^2 \leq \frac{\|\boldsymbol{B}_K^{-1}\|^2}{(1 - \gamma_0)^2} \left|\hat{\nabla}\hat{q})(\hat{\boldsymbol{x}})\right|^2 . \tag{2.80}$$

Then, we infer from (2.44) and (2.42) that

$$\frac{\int\limits_{\widehat{K}} \hat{b}(\hat{x}) \cdot \left|(\nabla q)\big(\boldsymbol{F}_K(\hat{\boldsymbol{x}})\big)\right|^2 \cdot |\det D\boldsymbol{F}_K(\hat{\boldsymbol{x}})|\, d\hat{\boldsymbol{x}}}{\int\limits_{\widehat{K}} \left|(\nabla q)\big(\boldsymbol{F}_K(\hat{\boldsymbol{x}})\big)\right|^2 \cdot |\det D\boldsymbol{F}_K(\hat{\boldsymbol{x}})|\, d\hat{\boldsymbol{x}}} \geq \frac{(1-\gamma_0)^{2+n}}{(1+\gamma_0)^{2+n}\|\boldsymbol{B}_K\|^2\,\|\boldsymbol{B}_K^{-1}\|^2} \, \frac{\int\limits_{\widehat{K}} \hat{b}(\hat{\boldsymbol{x}})\left|(\hat{\nabla}\hat{q}(\hat{\boldsymbol{x}})\right|^2 d\hat{\boldsymbol{x}}}{\int\limits_{\widehat{K}} \left|(\hat{\nabla}\hat{q}(\hat{\boldsymbol{x}})\right|^2 d\hat{\boldsymbol{x}}}\,. \tag{2.81}$$

Collecting (2.79),(2.81) and (2.42) implies

$$\big(\mathrm{div}(\varepsilon \boldsymbol{v}), q\big)_K \geq C_1 \varepsilon_0 \left(\frac{1-\gamma_0}{1+\gamma_0}\right)^{2+n} (1-\gamma_0)\, |q|_{1,K}^2\,. \tag{2.82}$$

Hereby $C_1 > 0$ is the constant satisfying

$$\|\sqrt{\hat{b}}\,\hat{\nabla}\hat{q}\|_{0,\widehat{K}} \geq \sqrt{C_1}\,\|\hat{\nabla}\hat{q}\|_{0,\widehat{K}}$$

due to the equivalence of the norms $\|\sqrt{\hat{b}}\,\hat{\nabla}(\cdot)\|_{0,\widehat{K}}$ and $\|\hat{\nabla}(\cdot)\|_{0,\widehat{K}}$ on the finite dimensional factor space $\widehat{P}_{k-1}/\mathbb{R}$. Next, we deduce from $\boldsymbol{v} = \hat{\boldsymbol{v}} \circ \boldsymbol{F}_K^{-1}$ and (2.77)

$$\hat{\boldsymbol{v}}(\hat{\boldsymbol{x}}) = -\hat{b}(\hat{\boldsymbol{x}}) \cdot \left\{\boldsymbol{I} + \boldsymbol{B}_K^{-1} D\boldsymbol{G}_K(\hat{\boldsymbol{x}})\right\}^T \cdot (\nabla q)\big(\boldsymbol{F}_K(\hat{\boldsymbol{x}})\big)\,,$$

and therefore we obtain with $\hat{b} \leq 1$ the estimate

$$\|\boldsymbol{v}\|_{0,K} \leq (1+\gamma_0)\, |q|_{1,K}\,. \tag{2.83}$$

Employing twice transformation rule, (2.44) and

$$(\nabla \boldsymbol{v})\big(\boldsymbol{F}_K(\hat{\boldsymbol{x}})\big) = \left\{\boldsymbol{I} + \boldsymbol{B}_K^{-1} D\boldsymbol{G}_K(\hat{\boldsymbol{x}})\right\}^{-T} \boldsymbol{B}_K^{-T} \cdot (\hat{\nabla}\hat{\boldsymbol{v}})(\hat{\boldsymbol{x}})\,,$$

we get

$$\begin{aligned} |\boldsymbol{v}|_{1,K}^2 &\leq (1-\gamma_0)^{-2}\, \|\boldsymbol{B}_K^{-T}\|^2 \int\limits_{\widehat{K}} (\hat{\nabla}\hat{\boldsymbol{v}})(\hat{\boldsymbol{x}}) : (\hat{\nabla}\hat{\boldsymbol{v}})(\hat{\boldsymbol{x}})\ |\det \boldsymbol{F}_K(\hat{\boldsymbol{x}})|\, d\hat{\boldsymbol{x}} \\ &\leq n!(1+\gamma_0)^n\, |S_K|\, (1-\gamma_0)^{-2}\, \|\boldsymbol{B}_K^{-T}\|^2\, |\hat{\boldsymbol{v}}|_{1,\widehat{K}}^2\,. \end{aligned}$$

From the fact that the norms $|\cdot|_{1,\widehat{K}}$ and $\|\cdot\|_{0,\widehat{K}}$ are equivalent on the finite dimensional space $\widehat{\boldsymbol{Q}}_k \cap \boldsymbol{H}_0^1(\widehat{K})$ follows the existence of a constant $C_2 > 0$ such that $|\hat{\boldsymbol{v}}|_{1,\widehat{K}} \leq \sqrt{C_2}\|\hat{\boldsymbol{v}}\|_{0,\widehat{K}}$. Then, using again transformation rule, (2.43) and (2.42), we get

$$|\boldsymbol{v}|_{1,K} \leq C_2 \left(\frac{1+\gamma_0}{1-\gamma_0}\right)^{n/2} (1-\gamma_0)^{-1}\, h_{S_K}^{-1}\, \|\boldsymbol{v}\|_{0,K} \qquad \forall\, \boldsymbol{v} \in \boldsymbol{X}_h(K)\,. \tag{2.84}$$

Analogously, we can state

$$\|q\|_{0,K} \le C_3 \left(\frac{1+\gamma_0}{1-\gamma_0}\right)^{n/2} (1+\gamma_0)\, h_{S_K}\, |q|_{1,K} \qquad \forall\, q \in M_h(K)\,, \tag{2.85}$$

where the constant $C_3 > 0$ satisfies

$$\|\hat{q}\|_{0,\widehat{K}} \le C_3 |\hat{q}|_{1,\widehat{K}} \quad \forall\, \hat{q} \in \widehat{P}_{k-1}/\mathbb{R}$$

due to the fact that the norms $\|\cdot\|_{0,\widehat{K}}$ and $|\cdot|_{1,\widehat{K}}$ are equivalent on the finite dimensional factor space $\hat{P}_{k-1}/\mathbb{R}$. Finally, using (2.82)-(2.85) we state the estimate

$$\begin{aligned}
\left(\operatorname{div}(\varepsilon \boldsymbol{v}), q\right)_K &\ge C_1 \varepsilon_0 \left(\frac{1-\gamma_0}{1+\gamma_0}\right)^{2+n} (1-\gamma_0)\, |q|^2_{1,K} \\
&\ge C_1 \varepsilon_0 \left(\frac{1-\gamma_0}{1+\gamma_0}\right)^{2+n} \frac{1-\gamma_0}{1+\gamma_0} \|\boldsymbol{v}\|_{0,K}\, |q|_{1,K} \\
&\ge C_1 C_2 C_3 \varepsilon_0 \left(\frac{1-\gamma_0}{1+\gamma_0}\right)^2 |\boldsymbol{v}|_{1,K}\, \|q\|_{0,K}
\end{aligned}$$

from which we conclude immediately the local inf-sup (2.50) condition with

$$\lambda = C_1 C_2 C_3 \varepsilon_0 \left(\frac{1-\gamma_0}{1+\gamma_0}\right)^2 .$$

□

Remark 2.21 *The inf-sup constant in (2.48) satisfies*

$$\gamma = \frac{\min\left\{\frac{1}{2}, \frac{\gamma_\varepsilon^2}{2 M_b C^2}\right\}}{\left\{\frac{1}{C_1^2 C_2^2 C_3^2 \varepsilon_0^2}\left(\frac{1+\gamma_0}{1-\gamma_0}\right)^4 + \left(\frac{\gamma_\varepsilon}{C M_b}\right)^2\right\}^{1/2}} \tag{2.86}$$

due to (2.62) and Remark 2.17. The constant $C > 0$ in the above relation corresponds to the stability constant in (2.65).

Combining results of Theorem 2.20 and Lemma 2.11, we deduce

Theorem 2.22 *Let $\mathcal{T}_h$ be shape regular. Then, the discrete problem (2.47) has a unique solution $(\boldsymbol{u}_h, p_h)$ in the space of mapped (Q_k, P_{k-1}^{disc}).*

2.3.3 Solvability of nonlinear discrete saddle point problem

Let $\boldsymbol{g}_h^*$ be the extension of $\boldsymbol{g}_h$ such that

$$\boldsymbol{g}_h^* = \boldsymbol{r}_h \boldsymbol{g}_\mu \quad \text{and} \quad \boldsymbol{g}_h^*|_\Gamma = (\boldsymbol{r}_h \boldsymbol{g}_\mu)|_\Gamma$$

where $\boldsymbol{r}_h : \boldsymbol{W} \to \boldsymbol{W}_h$ denotes the special interpolation operator which satisfies the modified divergence constraint in discrete sense and has the usual approximation properties. We consider mapped elements (Q_k, P_{k-1}^{disc}). The existence of the interpolation operator $\boldsymbol{r}_h : \boldsymbol{W} \to \boldsymbol{W}_h$ guarantees the following lemma.

Lemma 2.23 *There exists an interpolation operator $\boldsymbol{r}_h : \boldsymbol{W} \to \boldsymbol{W}_h$ such that*

$$|\boldsymbol{v} - \boldsymbol{r}_h \boldsymbol{v}|_1 \leq C h^{l-1} \|\boldsymbol{v}\|_l \quad \forall\, \boldsymbol{v} \in \boldsymbol{W} \cap \boldsymbol{H}^l(\Omega) \quad 1 \leq l \leq k+1\,, \tag{2.87}$$

and

$$b(\boldsymbol{r}_h \boldsymbol{v}, q_h) = 0 \quad \forall\, q_h \in M_h \quad \forall\, \boldsymbol{v} \in \boldsymbol{W}\,. \tag{2.88}$$

Proof. Let $B_h : \boldsymbol{W}_h \to M_h'$ be the linear continuous operator defined by

$$\langle B_h \boldsymbol{v}_h, q_h \rangle_{M_h} := b(\boldsymbol{v}_h, q_h) \qquad \forall\, \boldsymbol{v}_h \in \boldsymbol{W}_h \quad \forall\, q_h \in M_h\,.$$

After [30, Lemma 4.1, §4, Chapter I], the operator B_h is an isomorphism from $\boldsymbol{W}_h^\perp$ onto M_h' with

$$\gamma\, |\boldsymbol{v}_h|_1 \leq \|B_h \boldsymbol{v}_h\|_{M_h'} \qquad \forall\, \boldsymbol{v}_h \in \boldsymbol{W}_h^\perp \tag{2.89}$$

if and only if the inf-sup condition (2.48) holds. Here, M_h' denotes the dual space of M_h and $\boldsymbol{W}_h^\perp$ is L^2 orthogonal complement of $\boldsymbol{W}_h$ in $\boldsymbol{X}_{h0}$. Then, for each $\boldsymbol{v} \in \boldsymbol{W}$ there exists a unique $\boldsymbol{z}_h(\boldsymbol{v}) \in \boldsymbol{W}_h^\perp$ such that

$$\langle B_h \boldsymbol{z}_h(\boldsymbol{v}_h), q_h \rangle_{M_h} = b(\boldsymbol{z}_h(\boldsymbol{v}), q_h) = -b(\boldsymbol{i}_h \boldsymbol{v}, q_h) \quad \forall\, q_h \in M_h \tag{2.90}$$

where $\boldsymbol{i}_h : \boldsymbol{H}^1(\Omega) \to \boldsymbol{X}_{h0}$ is the interpolation operator of Scott–Zhang type with the usual stability and approximation properties (2.66) and (2.67). Thus, for all $\boldsymbol{v} \in \boldsymbol{W}$ holds

$$b(\boldsymbol{z}_h(\boldsymbol{v}), q_h) = -b(\boldsymbol{i}_h \boldsymbol{v} - \boldsymbol{v}, q_h) \quad \forall\, q_h \in M_h\,, \tag{2.91}$$

and according to (2.89) we obtain

$$|\boldsymbol{z}_h(\boldsymbol{v})|_1 \leq \frac{C_\varepsilon}{\gamma} |\boldsymbol{i}_h \boldsymbol{v} - \boldsymbol{v}|_1\,, \tag{2.92}$$

Taking

$$\boldsymbol{r}_h \boldsymbol{v} = \boldsymbol{i}_h \boldsymbol{v} + \boldsymbol{z}_h(\boldsymbol{v})$$

and using (2.67), we obtain

$$|\boldsymbol{v} - \boldsymbol{r}_h \boldsymbol{v}|_1 \leq \left(1 + \frac{C_\varepsilon}{\gamma}\right) |\boldsymbol{v} - \boldsymbol{i}_h \boldsymbol{v}|_1 \leq C h^{l-1} \|\boldsymbol{v}\|_l$$

for all $\boldsymbol{v} \in \boldsymbol{W}$ and $1 \leq l \leq k+1$. The property (2.88) follows from the definition of $\boldsymbol{r}_h \boldsymbol{v}$ and identity (2.90). □

We can state discrete estimates.

Lemma 2.24 *Let $\boldsymbol{g}_\mu$ be the extension of $\boldsymbol{g}$ defined by (2.13) and $\boldsymbol{g}_h^* = \boldsymbol{r}_h \boldsymbol{g}_\mu$. For any $\delta > 0$ there exist sufficiently small parameters $h > 0$, $\mu > 0$ such that*

$$|d(\boldsymbol{u}_h + \boldsymbol{g}_h^*; \boldsymbol{g}_h^*, \boldsymbol{u}_h)| \leq \delta \, \|\beta\|_{0,\infty} \, |\boldsymbol{u}_h|_1 \big(|\boldsymbol{u}_h|_1 + \|\boldsymbol{g}\|_0\big) \qquad \forall \, \boldsymbol{u} \in \boldsymbol{X}_{h0} \,, \tag{2.93}$$

$$|\tilde{n}(\boldsymbol{u}_h, \boldsymbol{g}_h^*, \boldsymbol{u}_h)| \leq \delta \, |\boldsymbol{u}_h|_1^2 \qquad \forall \, \boldsymbol{u}_h \in \boldsymbol{X}_{h0} \,. \tag{2.94}$$

Proof. Let $\boldsymbol{u}_h \in \boldsymbol{X}_{h0}$. Employing stability of interpolation and extension operators

$$\|\boldsymbol{r}_h \boldsymbol{g}_\mu\|_0 \leq \|\boldsymbol{r}_h \boldsymbol{g}_\mu\|_1 \leq C \|\boldsymbol{g}_\mu\|_1 \leq C \|\boldsymbol{g}\|_0 \,, \tag{2.95}$$

implies

$$\begin{aligned} |d(\boldsymbol{u}_h + \boldsymbol{g}_h^*; \boldsymbol{g}_h^*, \boldsymbol{u}_h)| &\leq |d(\boldsymbol{u}_h + \boldsymbol{g}_h^*, \boldsymbol{g}_h^* - \boldsymbol{g}_\mu, \boldsymbol{u}_h)| + |d(\boldsymbol{u}_h + \boldsymbol{g}_h^*, \boldsymbol{g}_\mu, \boldsymbol{u}_h)| \\ &\leq \|\beta\|_{0,\infty} \, (\|\boldsymbol{u}_h\|_0 + \|\boldsymbol{g}\|_0) \, \|\boldsymbol{g}_h^* - \boldsymbol{g}_\mu\|_0 \, \|\boldsymbol{u}_h\|_0 \\ &\quad + \|\beta\|_{0,\infty} \, (\|\boldsymbol{u}_h\|_0 + \|\boldsymbol{g}\|_0) \, \|\boldsymbol{u}_h \cdot \boldsymbol{g}_\mu\|_0 \,. \end{aligned}$$

The bound of the interpolation error follows from the properties of extension $\boldsymbol{g}_\mu|_\Gamma = \boldsymbol{g}$

$$\|\boldsymbol{g}_h^* - \boldsymbol{g}_\mu\|_l \leq Ch |\boldsymbol{g}_\mu|_1 \leq Ch \|\boldsymbol{g}\|_0 \,, \quad l = 0, 1. \tag{2.96}$$

Then, according to Lemma 2.4 we can state that for each $\delta > 0$ there exists $\mu > 0$ such that

$$|d(\boldsymbol{u}_h + \boldsymbol{g}_h^*; \boldsymbol{g}_h^*, \boldsymbol{u}_h)| \leq \delta \, \|\beta\|_{0,\infty} \, |\boldsymbol{u}_h|_1 \, (\|\boldsymbol{u}_h\|_0 + \|\boldsymbol{g}\|_0)$$

holds for sufficiently small $h > 0$. For the trilinear term we have the estimate

$$|\tilde{n}(\boldsymbol{u}_h, \boldsymbol{g}_h^*, \boldsymbol{u}_h)| \leq |\tilde{n}(\boldsymbol{u}_h, \boldsymbol{g}_h^* - \boldsymbol{g}_\mu, \boldsymbol{u}_h)| + |\tilde{n}(\boldsymbol{u}_h, \boldsymbol{g}_\mu, \boldsymbol{u}_h)| \,. \tag{2.97}$$

Using (2.10), (2.96) and Poincaré inequality, we obtain for sufficiently small $h > 0$

$$|\tilde{n}(\boldsymbol{u}_h, \boldsymbol{g}_h^* - \boldsymbol{g}_\mu, \boldsymbol{u}_h)| \leq Ch \|\boldsymbol{g}\|_0 \, \|\boldsymbol{u}_h\|_1^2 \leq \delta \, \|\boldsymbol{g}\|_0 \, \|\boldsymbol{u}_h\|_1^2 \,. \tag{2.98}$$

Next, we have

$$\tilde{n}(\boldsymbol{u}_h, \boldsymbol{g}_\mu, \boldsymbol{u}_h) = \frac{1}{2} \big[n(\boldsymbol{u}_h, \boldsymbol{g}_\mu, \boldsymbol{u}_h) - n(\boldsymbol{u}_h, \boldsymbol{u}_h, \boldsymbol{g}_\mu)\big] \,. \tag{2.99}$$

The first term in the above square bracket gets small due to Lemma 2.4 and the smallness of the second one follows from the Cauchy-Schwarz inequality, (2.19) and Poincaré inequality

$$n(\boldsymbol{u}_h, \boldsymbol{u}_h, \boldsymbol{g}_\mu) \leq C \|\boldsymbol{u}_h\|_1 \|\boldsymbol{u}_h \cdot \boldsymbol{g}_\mu\|_0 < C\delta |\boldsymbol{u}_h|_1 \,. \tag{2.100}$$

Collecting (2.97)-(2.100), we deduce the estimate (2.94). □

Remark 2.25 *The extension $\boldsymbol{g}_h^*$ is used only for the sake of error analysis and does not need to be constructed during numerical calculations.*

Let $G_h : \boldsymbol{W}_h \to \boldsymbol{W}_h$ be the operator defined by

$$\begin{aligned}\left[G_h(\boldsymbol{w}_h), \boldsymbol{v}_h\right] :=& a(\boldsymbol{w}_h + \boldsymbol{g}_h^*, \boldsymbol{v}_h) + c(\boldsymbol{w}_h + \boldsymbol{g}_h^*, \boldsymbol{v}_h) - (\boldsymbol{f}, \boldsymbol{v_h}) \\ &+ \tilde{n}(\boldsymbol{w}_h + \boldsymbol{g}_h^*, \boldsymbol{w}_h + \boldsymbol{g}_h^*, \boldsymbol{v}_h) + d(\boldsymbol{w}_h + \boldsymbol{g}_h^*; \boldsymbol{w}_h + \boldsymbol{g}_h^*, \boldsymbol{v}_h)\,.\end{aligned} \tag{2.101}$$

Then, the discrete problem (2.46) can be rewritten as operator problem

Find $\boldsymbol{w}_h \in \boldsymbol{W}_h$ such that

$$\left[G_h(\boldsymbol{w}_h), \boldsymbol{v}_h\right] = 0 \quad \forall\, \boldsymbol{v}_h \in \boldsymbol{W}_h. \tag{2.102}$$

Now, we show that the discrete anologon of Lemma 2.5 holds.

Lemma 2.26 *The mapping G_h defined in (2.101) is continuous and there exists $r > 0$ s.t.*

$$\left[G_h(\boldsymbol{u}_h), \boldsymbol{u}_h\right] > 0 \quad \forall\, \boldsymbol{u}_h \in \boldsymbol{W}_h \quad \textit{with} \quad |\boldsymbol{u}_h|_1 = r. \tag{2.103}$$

Proof. The continuity of the operator G_h can be shown in the same way as in the proof of Lemma 2.5. From Lemma 2.24 we deduce by analogy to the proof of Lemma 2.5

$$\begin{aligned}\left[G(\boldsymbol{u}_h), \boldsymbol{u}_h\right] \geq& \left\{\frac{\varepsilon_0}{Re} - \delta(1 + \|\beta\|_{0,\infty})\right\} |\boldsymbol{u}_h|_1^2 \\ &- \left\{\frac{\varepsilon_1}{Re}\|\boldsymbol{g}\|_0 + C\frac{1}{Re}\|\alpha\|_{0,\infty}\|\boldsymbol{g}\|_0 + \delta\|\beta\|_{0,\infty}\|\boldsymbol{g}\|_0 + C\|\boldsymbol{g}\|_0^2 + C\|\boldsymbol{f}\|_0\right\}|\boldsymbol{u}_h|_1.\end{aligned}$$

The choice

$$0 < \delta < \delta_0 := \frac{\varepsilon_0}{Re}\left(1 + \|\beta\|_{0,\infty}\right)^{-1},$$

and $r > r_0$ with

$$r_0 := \frac{\frac{\varepsilon_1}{Re}\|\boldsymbol{g}\|_0 + \frac{1}{Re}C\|\alpha\|_{0,\infty}\|\boldsymbol{g}\|_0 + \delta\|\beta\|_{0,\infty}\|\boldsymbol{g}\|_0 + C\|\boldsymbol{g}\|_0^2 + C\|\boldsymbol{f}\|_0}{\frac{\varepsilon_0}{Re} - \delta(1 + \|\beta\|_{0,\infty})}, \tag{2.104}$$

yields the assertion (2.103). □

Now, we can state the main result concerning solvability of the discrete nonlinear problem (2.46).

Theorem 2.27 *The discrete nonlinear problem (2.46) has at least one solution belonging to the space of mapped (Q_k, P_{k-1}^{disc}) elements.*

Proof. Employing results of Lemma 2.26, the existence of the discrete velocity is a straightforward consequence of Lemma 2.6 applied to the operator G_h. The reconstruction of the discrete pressure can be proceeded in the same way as in the continuous problem. From the inf-sup condition satisfied due to Theorem 2.20 we deduce the existence of the discrete pressure. □

For sufficiently small data we can also show that the discrete solution is unique.

Theorem 2.28 *If $\|\boldsymbol{g}\|_0$, $\|\boldsymbol{f}\|_{-1}$ are sufficiently small, then the solution of the discrete problem (2.46) is unique.*

Proof. We follow the proof line of Theorem 2.10. Assume that $(\boldsymbol{u}_{h1}, p_{h1})$ and $(\boldsymbol{u}_{h2}, p_{h2})$ are two different solutions of (2.46). Then, we obtain

$$\begin{aligned}
0 &= \big[G(\boldsymbol{u}_{h1}) - G(\boldsymbol{u}_{h2}), \boldsymbol{u}_{h1} - \boldsymbol{u}_{h2}\big] \\
&= a(\boldsymbol{u}_{h1} - \boldsymbol{u}_{h2}, \boldsymbol{u}_{h1} - \boldsymbol{u}_{h2}) + c(\boldsymbol{u}_{h1} - \boldsymbol{u}_{h2}, \boldsymbol{u}_{h1} - \boldsymbol{u}_{h2}) - (\boldsymbol{f}, \boldsymbol{u}_{h1} - \boldsymbol{u}_{h2}) \\
&\quad + \tilde{n}(\boldsymbol{u}_{h1} + \boldsymbol{g}_h^*, \boldsymbol{u}_{h1} + \boldsymbol{g}_h^*, \boldsymbol{u}_{h1} - \boldsymbol{u}_{h2}) - \tilde{n}(\boldsymbol{u}_{h2} + \boldsymbol{g}_h^*, \boldsymbol{u}_{h2} + \boldsymbol{g}_h^*, \boldsymbol{u}_{h1} - \boldsymbol{u}_{h2}) \\
&\quad + (\beta|\boldsymbol{u}_{h1} + \boldsymbol{g}_h^*|(\boldsymbol{u}_{h1} + \boldsymbol{g}_h^*), \boldsymbol{u}_{h1} - \boldsymbol{u}_{h2}) \\
&\quad - \big(\beta|\boldsymbol{u}_{h2} + \boldsymbol{g}_h^*|(\boldsymbol{u}_{h2} + \boldsymbol{g}_h^*), \boldsymbol{u}_{h1} - \boldsymbol{u}_{h2}\big) \\
&\geq \frac{\varepsilon_0}{Re}\|\boldsymbol{u}_{h1} - \boldsymbol{u}_{h2}\|_1^2 - \|\boldsymbol{f}\|_{-1}\|\boldsymbol{u}_{h1} - \boldsymbol{u}_{h2}\|_1 \\
&\quad - |\tilde{n}(\boldsymbol{u}_{h1} - \boldsymbol{u}_{h2}, \boldsymbol{u}_{h2}, \boldsymbol{u}_{h1} - \boldsymbol{u}_{h2})| - |\tilde{n}(\boldsymbol{u}_{h1} - \boldsymbol{u}_{h2}, \boldsymbol{g}_h^*, \boldsymbol{u}_{h1} - \boldsymbol{u}_{h2})| \\
&\quad - \|\beta\|_{0,\infty}\left|\big(|\boldsymbol{u}_{h1} + \boldsymbol{g}_h^*|(\boldsymbol{u}_{h1} - \boldsymbol{u}_{h2}), \boldsymbol{u}_{h1} - \boldsymbol{u}_{h2}\big)\right| \\
&\quad - \|\beta\|_{0,\infty}\left|\big((|\boldsymbol{u}_{h1} + \boldsymbol{g}_h^*| - |\boldsymbol{u}_{h2} + \boldsymbol{g}_h^*|)(\boldsymbol{u}_{h2} + \boldsymbol{g}_h^*), \boldsymbol{u}_{h1} - \boldsymbol{u}_{h2}\big)\right| .
\end{aligned} \tag{2.105}$$

From stability estimate (2.95) and Cauchy-Schwarz inequality we deduce

$$\left|\big(|\boldsymbol{u}_{h1} + \boldsymbol{g}_h^*|\,(\boldsymbol{u}_{h1} - \boldsymbol{u}_{h2}), \boldsymbol{u}_{h1} - \boldsymbol{u}_{h2}\big)\right| \leq C\left\{\|\boldsymbol{u}_{h1}\|_0 + \|\boldsymbol{g}\|_0\right\}\|\boldsymbol{u}_{h1} - \boldsymbol{u}_{h2}\|_1^2, \tag{2.106}$$

$$\begin{aligned}
&\left|\big((|\boldsymbol{u}_{h1} + \boldsymbol{g}_h^*| - |\boldsymbol{u}_{h2} + \boldsymbol{g}_h^*|)(\boldsymbol{u}_{h2} + \boldsymbol{g}_h^*), \boldsymbol{u}_{h1} - \boldsymbol{u}_{h2}\big)\right| \\
&\qquad \leq C\left\{\|\boldsymbol{u}_{h2}\|_0 + \|\boldsymbol{g}\|_0\right\}\|\boldsymbol{u}_{h1} - \boldsymbol{u}_{h2}\|_1^2 .
\end{aligned} \tag{2.107}$$

According to (2.10) we have

$$|\tilde{n}(\boldsymbol{u}_{h1} - \boldsymbol{u}_{h2}, \boldsymbol{u}_{h2}, \boldsymbol{u}_{h1} - \boldsymbol{u}_{h2})| \leq C\|\boldsymbol{u}_{h2}\|_1\|\boldsymbol{u}_{h1} - \boldsymbol{u}_{h2}\|_1^2, \tag{2.108}$$

and by (2.94) we can find μ such that

$$|\tilde{n}(\boldsymbol{u}_{h1} - \boldsymbol{u}_{h2}, \boldsymbol{g}_h^*, \boldsymbol{u}_{h1} - \boldsymbol{u}_{h2})| \leq \frac{\varepsilon_0}{4Re}\|\boldsymbol{u}_{h1} - \boldsymbol{u}_{h2}\|_1^2. \tag{2.109}$$

Employing stability estimate (2.95) we find by analogy to the continuous case the upper bound for the discrete solution

$$\|\boldsymbol{u}_h\|_1 \leq \frac{\|\boldsymbol{f}\|_{-1} + C_1\|\boldsymbol{g}\|_0 + C_2\|\boldsymbol{g}\|_0^2}{\dfrac{\varepsilon_0}{Re} - C_3\|\beta\|_{0,\infty}\|\boldsymbol{g}\|_0} := C\big(\|\boldsymbol{g}\|_0, \|\boldsymbol{f}\|_{-1}\big) \quad \forall\, \boldsymbol{u}_h \in \boldsymbol{W}_h . \tag{2.110}$$

Combining (2.105)-(2.110) and using the inequality

$$\|\boldsymbol{f}\|_{-1}\|\boldsymbol{u}_{h1}-u_{h2}\|_1 \le \frac{\varepsilon_0}{4Re}\|\boldsymbol{u}_{h1}-\boldsymbol{u}_{h2}\|_1^2 + \frac{2Re}{\varepsilon_0}\|\boldsymbol{f}\|_{-1}^2\,,$$

we state

$$\begin{aligned} 0 \ge \frac{\varepsilon_0}{2Re}\|\boldsymbol{u}_{h1}-\boldsymbol{u}_{h2}\|_1^2 &- \frac{2Re}{\varepsilon_0}\|\boldsymbol{f}\|_{-1}^2 - C\big(\|\boldsymbol{g}_0\|_0, \|\boldsymbol{f}\|_{-1}\big)\,\|\beta\|_{0,\infty}\|\boldsymbol{u}_{h1}-\boldsymbol{u}_{h2}\|_1^2 \\ - \frac{\varepsilon_0}{4Re}\|\boldsymbol{u}_{h1}-\boldsymbol{u}_{h2}\|_1^2 &- C\big(\|\boldsymbol{g}\|_0, \|\boldsymbol{f}\|_{-1}\big)\|\boldsymbol{u}_{h1}-\boldsymbol{u}_{h2}\|_1^2\,. \end{aligned} \tag{2.111}$$

The upper bound $C(\|\boldsymbol{g}\|_0, \|\boldsymbol{f}\|_1)$ in (2.110) gets small for sufficiently small $\|\boldsymbol{g}\|_0$, $\|\boldsymbol{f}\|_1$. Consequently, the right hand side of (2.111) is nonnegative. This implies $\boldsymbol{u}_{h1} = \boldsymbol{u}_{h2}$. From Theorem 2.20 follows immediately $p_{h1} - p_{h2} = const.$ □

2.3.4 Error estimates

Let $\boldsymbol{\xi}_h \in \boldsymbol{W}_h$. Combining equations (2.8) and (2.46) we deduce

$$\begin{aligned} 0 &= a(\boldsymbol{u}_h-\boldsymbol{u},\boldsymbol{\xi}_h) + c(\boldsymbol{u}_h-\boldsymbol{u},\boldsymbol{\xi}_h) \\ &\quad + \tilde{n}(\boldsymbol{u}_h,\boldsymbol{u}_h,\boldsymbol{\xi}_h) - n(\boldsymbol{u},\boldsymbol{u},\boldsymbol{\xi}_h) \\ &\quad + d(\boldsymbol{u}_h;\boldsymbol{u}_h,\boldsymbol{\xi}_h) - d(\boldsymbol{u};\boldsymbol{u},\boldsymbol{\xi}_h) \\ &\quad - b(\boldsymbol{\xi}_h, p_h - p)\,. \end{aligned}$$

Choosing $\boldsymbol{\xi}_h = \boldsymbol{u}_h - \boldsymbol{v}_h$ with $\boldsymbol{v}_h$ such that $\boldsymbol{v}_h - \boldsymbol{g}_h^* \in \boldsymbol{W}_h$, we obtain the identity

$$\begin{aligned} a(\boldsymbol{\xi}_h,\boldsymbol{\xi}_h) + c(\boldsymbol{\xi}_h,\boldsymbol{\xi}_h) &= a(\boldsymbol{u}-\boldsymbol{v}_h,\boldsymbol{\xi}_h) + c(\boldsymbol{u}-\boldsymbol{v}_h,\boldsymbol{\xi}_h) \\ &\quad + \big\{n(\boldsymbol{u},\boldsymbol{u},\boldsymbol{\xi}_h) - \tilde{n}(\boldsymbol{v}_h,\boldsymbol{v}_h,\boldsymbol{\xi}_h)\big\} \\ &\quad + \big\{\tilde{n}(\boldsymbol{v}_h,\boldsymbol{u}_h,\boldsymbol{\xi}_h) - \tilde{n}(\boldsymbol{u}_h,\boldsymbol{u}_h,\boldsymbol{\xi}_h)\big\} \\ &\quad - \tilde{n}(\boldsymbol{v}_h,\boldsymbol{\xi}_h,\boldsymbol{\xi}_h) \\ &\quad + \big\{d(\boldsymbol{u};\boldsymbol{u},\boldsymbol{\xi}_h) - d(\boldsymbol{u}_h;\boldsymbol{u}_h,\boldsymbol{\xi}_h)\big\} \\ &\quad + b(\boldsymbol{\xi}_h, p_h - p)\,. \end{aligned} \tag{2.112}$$

The first two terms on the right hand side we estimate by using Cauchy-Schwarz inequality

$$a(\boldsymbol{u}-\boldsymbol{v}_h,\boldsymbol{\xi}_h) \le \|\varepsilon\|_{0,\infty}\,|\boldsymbol{u}-\boldsymbol{v}_h|_1\,|\boldsymbol{\xi}_h|_1\,, \tag{2.113}$$

$$c(\boldsymbol{u}-\boldsymbol{v}_h,\boldsymbol{\xi}_h) \le C_\varepsilon\,\|\boldsymbol{u}-\boldsymbol{v}_h\|_1\,\|\boldsymbol{\xi}_h\|_1\,. \tag{2.114}$$

We denote by $j_h : L_0^2(\Omega) \cap H^k(\Omega) \to M_h$ the finite element interpolation operator for the pressure. Applying Hölder inequality and the optimal interpolation estimate for the pressure, implies

$$\begin{aligned} b(\boldsymbol{\xi}_h, p_h - p) &= b(\boldsymbol{\xi}_h, j_h p - p) \\ &\leq (|\varepsilon|_{1,3} + \|\varepsilon\|_{0,\infty}) \|\boldsymbol{\xi}_h\|_1 \|j_h p - p\|_0 \\ &\leq C_\varepsilon h^k \|\boldsymbol{\xi}_h\|_1 |p|_k . \end{aligned} \tag{2.115}$$

Employing a priori estimate and the continuity of the trilinear form $n(\cdot,\cdot,\cdot)$, we obtain for the difference in the first brace

$$\begin{aligned} & n(\boldsymbol{u}, \boldsymbol{u}, \boldsymbol{\xi}_h) - \tilde{n}(\boldsymbol{v}_h, \boldsymbol{v}_h, \boldsymbol{\xi}_h) \\ & = n(\boldsymbol{u}, \boldsymbol{u} - \boldsymbol{v}_h, \boldsymbol{\xi}_h) + n(\boldsymbol{u} - \boldsymbol{v}_h, \boldsymbol{v}_h - \boldsymbol{u}, \boldsymbol{\xi}_h) + n(\boldsymbol{u} - \boldsymbol{v}_h, \boldsymbol{u}, \boldsymbol{\xi}_h) \\ & \leq C_\varepsilon \|\boldsymbol{u}\|_1 \|\boldsymbol{u} - \boldsymbol{v}_h\|_1 \|\boldsymbol{\xi}_h\|_1 + C_\varepsilon \|\boldsymbol{u} - \boldsymbol{v}_h\|_1^2 \|\boldsymbol{\xi}_h\|_1 . \end{aligned} \tag{2.116}$$

From the a priori estimate for the discrete velocity we deduce the bound for the difference in the second brace

$$\begin{aligned} & \tilde{n}(\boldsymbol{v}_h, \boldsymbol{u}_h, \boldsymbol{\xi}_h) - \tilde{n}(\boldsymbol{u}_h, \boldsymbol{u}_h, \boldsymbol{\xi}_h) = -\tilde{n}(\boldsymbol{\xi}_h, \boldsymbol{u}_h, \boldsymbol{\xi}_h) \\ & \leq C(\|\boldsymbol{g}\|_0, \|\boldsymbol{f}\|_{-1}) \|\boldsymbol{\xi}_h\|_1^2 . \end{aligned} \tag{2.117}$$

From the definition of $\tilde{n}(\cdot,\cdot,\cdot)$ follows immediately

$$\tilde{n}(\boldsymbol{v}_h, \boldsymbol{\xi}_h, \boldsymbol{\xi}_h) = 0 . \tag{2.118}$$

We rewrite the third difference in the brace

$$\begin{aligned} & d(\boldsymbol{u}; \boldsymbol{u}, \boldsymbol{\xi}_h) - d(\boldsymbol{u}_h; \boldsymbol{u}_h, \boldsymbol{\xi}_h) \\ & = d(\boldsymbol{u}; \boldsymbol{u} - \boldsymbol{v}_h, \boldsymbol{\xi}_h) + d(\boldsymbol{u}; \boldsymbol{v}_h - \boldsymbol{u}_h, \boldsymbol{\xi}_h) \\ & \quad + d(\boldsymbol{u}; \boldsymbol{u}_h, \boldsymbol{\xi}_h) - d(\boldsymbol{v}_h; \boldsymbol{u}_h, \boldsymbol{\xi}_h) \\ & \quad + d(\boldsymbol{v}_h; \boldsymbol{u}_h, \boldsymbol{\xi}_h) - d(\boldsymbol{u}_h; \boldsymbol{u}_h, \boldsymbol{\xi}_h) . \end{aligned} \tag{2.119}$$

Employing continuity of $d(\cdot;\cdot,\cdot)$, a priori estimate (2.110) and the inverse triangle inequality, implies

$$\begin{aligned} & d(\boldsymbol{u}; \boldsymbol{u}, \boldsymbol{\xi}_h) - d(\boldsymbol{u}_h; \boldsymbol{u}_h, \boldsymbol{\xi}_h) \\ & \leq C_\varepsilon \|\boldsymbol{u}\|_1 \|\boldsymbol{u} - \boldsymbol{v}_h\|_1 \|\boldsymbol{\xi}_h\|_1 \\ & \quad + C(\|\boldsymbol{g}\|_0, \|\boldsymbol{f}\|_{-1}) \|\boldsymbol{\xi}_h\|_1^2 \\ & \quad + C_\varepsilon \|\boldsymbol{u} - \boldsymbol{v}_h\|_1 \|\boldsymbol{w}\|_1 \\ & \quad + C(\|\boldsymbol{g}\|_0, \|\boldsymbol{f}\|_{-1}) \|\boldsymbol{\xi}_h\|_1^2 . \end{aligned} \tag{2.120}$$

Collecting (2.113)-(2.120) and using Poincaré inequality, we obtain the estimate

$$\begin{aligned} \left\{ \frac{\varepsilon_0}{Re} - C(\|\boldsymbol{g}\|_0, \|\boldsymbol{f}\|_{-1}) \right\} |\boldsymbol{\xi}_h|_1 \leq \|\varepsilon\|_{0,\infty} |\boldsymbol{u} - \boldsymbol{v}_h|_1 + C\|\boldsymbol{u} - \boldsymbol{v}_h\|_1 \\ + C\|\boldsymbol{u}\|_1 \|\boldsymbol{u} - \boldsymbol{v}_h\|_1 + C\|\boldsymbol{u} - \boldsymbol{v}_h\|_1^2 . \end{aligned} \tag{2.121}$$

Now, we are able to state the error estimates for finite element discretisation of nonlinear reactor flow problem.

Theorem 2.29 *Let $(\boldsymbol{u}, p) \in \boldsymbol{X} \times M$ be solution of the weak problem (2.8). The discrete solution $(\boldsymbol{u}_h, p_h) \in \boldsymbol{X}_h \times M_h$ belonging to the space of mapped (Q_k, P_{k-1}^{disc}) elements satisfies the following error estimates*

$$|\boldsymbol{u} - \boldsymbol{u}_h|_1 \leq Ch^k \quad \textit{and} \quad \|p - p_h\|_0 \leq Ch^k\,, \tag{2.122}$$

provided that $Re > 0$, $\|\boldsymbol{g}\|_0$, $\|\boldsymbol{f}\|_{-1}$ and $h > 0$ are sufficiently small.

Proof. Let $\boldsymbol{u} = \boldsymbol{w} + \boldsymbol{g}_\mu \in \boldsymbol{X}$ and $p \in M$ be the weak solution of problem (2.46) where $\boldsymbol{w} \in \boldsymbol{W}$ solves (2.24) and $\boldsymbol{g}_\mu$ is the extension defined by (2.13). We consider the solution of discrete problem (2.46) in the mapped space pair (Q_k, P_{k-1}^{disc}). We set as usual $\boldsymbol{u}_h = \boldsymbol{w}_h + \boldsymbol{g}_h^*$ where $\boldsymbol{w}_h \in \boldsymbol{W}_h$ solves (2.102) and $\boldsymbol{g}_h^* = \boldsymbol{r}_h \boldsymbol{g}_\mu$ as in Lemma 2.24. The weak solution can be interpolated by $\boldsymbol{r}_h \boldsymbol{w} + \boldsymbol{g}_h^*$ where $\boldsymbol{r}_h \boldsymbol{w}$ is the finite element interpolant established in Lemma 2.23. Setting $\boldsymbol{\xi}_h = \boldsymbol{w}_h - \boldsymbol{r}_h \boldsymbol{w} \in \boldsymbol{W}_h$ and using triangle inequality, implies

$$|\boldsymbol{u} - \boldsymbol{u}_h|_1 \leq |\boldsymbol{w} - \boldsymbol{r}_h \boldsymbol{w}|_1 + |\boldsymbol{g}_\mu - \boldsymbol{r}_h \boldsymbol{g}_\mu|_1 + |\boldsymbol{\xi}_h|_1\,.$$

Taking $\boldsymbol{v}_h = \boldsymbol{r}_h \boldsymbol{w} + \boldsymbol{g}_h^*$ in (2.121) and employing interpolation estimates (2.87) and (2.67), we obtain for sufficiently small data and mesh parameter $h > 0$

$$|\boldsymbol{\xi}_h|_1 \leq Ch^k\,. \tag{2.123}$$

Consequently, the error bound for velocity holds. In the next stage we estimate the pressure error. To this end, we taking $q_h = p_h - j_h p$. The bound for the pressure error can be derived from the following identity

$$\begin{aligned} -b(\boldsymbol{\xi}_h, q_h) &= -b(\boldsymbol{\xi}_h, p - j_h p) + a(\boldsymbol{u} - \boldsymbol{u}_h, \boldsymbol{\xi}_h) + c(\boldsymbol{u} - \boldsymbol{u}_h, \boldsymbol{\xi}_h) \\ &\quad + n(\boldsymbol{u}, \boldsymbol{u}, \boldsymbol{\xi}_h) - \tilde{n}(\boldsymbol{u}_h, \boldsymbol{u}_h, \boldsymbol{\xi}_h) \\ &\quad + d(\boldsymbol{u}; \boldsymbol{u}, \boldsymbol{\xi}_h) - d(\boldsymbol{u}_h; \boldsymbol{u}_h, \boldsymbol{\xi}_h)\,. \end{aligned}$$

Again, using the continuity of bilinear form $b(\cdot, \cdot)$, the interpolation error estimate for the pressure, exploiting bounds in (2.113)-(2.120), applying the discrete inf-sup condition (2.48), we get from (2.123)

$$\|q_h\|_0 \leq \frac{C}{\gamma} h^k + \frac{C}{\gamma} h^k |\boldsymbol{\xi}_h|_1 \leq Ch^k\,.$$

Taking $q_h = p_h - j_h p$ and invoking interpolation error estimate, we deduce from the triangle inequality the error bound

$$\|p - p_h\|_0 \leq \|p - j_h p\|_0 + \|q_h\|_0 \leq Ch^k$$

which completes the proof. □

Remark 2.30 *The construction of $\boldsymbol{r}_h \boldsymbol{g}_\mu$ is not required during the numerical computations. Theoretical results stated in [26, 33] and our numerical tests indicate that the restriction $(\boldsymbol{r}_h \boldsymbol{g}_\mu)|_\Gamma$ can be replaced by the usual Lagrange interpolant of $\boldsymbol{g}$.*

Remark 2.31 *Results stated in [8] can be applied to our model problem. The error analysis done for the nonsymetric saddle point problem involves more sophisticated techniques and gives not optimal order of convergence since the order of L^2 interpolation error for the used pressure elements is suboptimal.*

2.4 Numerical results

2.4.1 Problem with smooth solution

We start our numerical investigations solving a two dimensional problem which is posed on the domain $\Omega = (0,1)^2$. We apply stable (Q_2, P_1^{disc}) and (Q_3, P_2^{disc}) elements on cartesian meshes. The coarse mesh consists of 2×2 squares and will be uniformly refined. The

Table 2.1: Total number of degrees of freedom (dof) for velocity and pressure.

level	dofs			
	Q_2	P_1^{disc}	Q_3	P_2^{disc}
0	50	12	98	24
1	162	48	338	96
2	578	192	1,250	384
3	2,178	768	4,802	1,536
4	8,450	3,072	18,818	6,144
5	33,282	12,288	74,498	24,576
6	132,098	49,152	296,450	98,304

corresponding numbers of degrees of freedom for velocity and pressure are shown in Table 2.1. First, we report results for Stokes-like problem

Find $(\boldsymbol{u}, p) \in \boldsymbol{X} \times M$, with $\boldsymbol{u}|_\Gamma = \boldsymbol{g}$, such that

$$a(\boldsymbol{u}, \boldsymbol{v}) - b(\boldsymbol{v}, p) + b(\boldsymbol{u}, q) = (\boldsymbol{f}, \boldsymbol{v}) \quad \forall\, (\boldsymbol{v}, q) \in \boldsymbol{V}\,. \tag{2.124}$$

The right hand side $\boldsymbol{f}$ and boundary condition $\boldsymbol{g}$ are chosen such that

$$\begin{cases} \boldsymbol{u}(x,y) &= \dfrac{1}{\varepsilon(x,y)} \begin{pmatrix} \sin(\pi x)\sin(\pi y) \\ \cos(\pi x)\cos(\pi y) \end{pmatrix} \\ p(x,y) &= 2\cos(\pi x)\,\sin(\pi y) \end{cases} \tag{2.125}$$

is the solution of the problem. The fictitious porosity distribution is defined as

$$\varepsilon(x,y) = 1 - \frac{1}{2}\sin(\pi x)\,\sin(\pi y)\,. \tag{2.126}$$

The velocity and pressure error measured with H^1 seminorm and L^2 norm, respectively, are presented in Table 2.2 and Table 2.3. Additionally, we report the velocity error measured with L_2 norm.

Table 2.2: Stokes-like problem: velocity and pressure errors with rates of convergence for (Q_2, P_1^{disc}) elements

level	$\|\boldsymbol{u}-\boldsymbol{u}_h\|_1$	rate	$\|p-p_h\|_0$	rate	$\|\boldsymbol{u}-\boldsymbol{u}_h\|_0$	rate
0	1.070e+0		3.143e−1		7.765e−2	
1	2.777e−1	1.947	6.969e−2	2.173	1.038e−2	2.903
2	6.528e−2	2.088	1.568e−2	2.153	1.243e−3	3.062
3	1.642e−2	1.991	3.828e−3	2.034	1.579e−4	2.977
4	4.113e−3	1.998	9.520e−4	2.008	1.981e−5	2.994
5	1.029e−3	1.999	2.377e−4	2.002	2.479e−6	2.999
6	2.572e−4	2.000	5.940e−5	2.000	3.100e−7	3.000

Table 2.3: Stokes-like problem: Velocity and pressure errors with rates of convergence for (Q_3, P_2^{disc}) elements.

level	$\|\boldsymbol{u}-\boldsymbol{u}_h\|_1$	rate	$\|p-p_h\|_0$	rate	$\|\boldsymbol{u}-\boldsymbol{u}_h\|_0$	rate
0	2.539e−1		6.894e−2		1.349e−2	
1	3.051e−2	3.057	8.061e−3	3.096	7.720e−4	4.127
2	5.639e−3	2.436	9.939e−4	3.020	7.410e−5	3.381
3	7.161e−4	2.977	1.219e−4	3.028	4.713e−6	3.975
4	8.983e−5	2.995	1.511e−5	3.012	2.957e−7	3.994
5	1.124e−5	2.999	1.881e−6	3.005	1.850e−8	3.998
6	1.405e−6	3.000	2.348e−7	3.003	1.157e−9	4.000

Next, we report in the same way results for Brinkman–Forcheimer problem with Reynolds number $Re = 1$. We stop the fixed point iteration (2.47) if two successive solutions of algebraic systems differ less than $1e-8$ with respect to the Euclidian norm. In Table 2.4 and 2.5 results are presented for (Q_2, P_1^{disc}) and (Q_3, P_2^{disc}) elements, respectively. The calculated rates of convergence are in good agreement with theoretical results from Section 2.3. The asymptotic behaviour of total error $|\boldsymbol{u}-\boldsymbol{u}_h|_1 + \|p-p_h\|_0$ is shown in Figures 2.1 and 2.2 for Stokes-like and Brinkman problem, respectively.

Table 2.4: Velocity and pressure errors with rates of convergence for (Q_2, P_1^{disc}) elements.

level	$\|\boldsymbol{u}-\boldsymbol{u}_h\|_1$	rate	$\|p-p_h\|_0$	rate	$\|\boldsymbol{u}-\boldsymbol{u}_h\|_0$	rate
0	1.114e+0		1.886e+0		6.949e−2	
1	2.799e−1	1.992	1.257e−1	3.907	9.508e−3	2.870
2	6.532e−2	2.200	1.663e−2	2.918	1.219e−3	2.964
3	1.642e−2	1.991	3.845e−3	2.113	1.572e−4	2.955
4	4.113e−3	1.998	9.523e−4	2.013	1.979e−5	2.989
5	1.029e−3	1.999	2.377e−4	2.002	2.479e−6	2.997
6	2.572e−4	2.000	5.940e−5	2.001	3.100e−7	2.999

Table 2.5: Velocity and pressure errors with rates of convergence for (Q_3, P_2^{disc}) elements.

level	$\|\boldsymbol{u}-\boldsymbol{u}_h\|_1$	rate	$\|p-p_h\|_0$	rate	$\|\boldsymbol{u}-\boldsymbol{u}_h\|_0$	rate
0	2.512e−1		8.567e−2		1.277e−2	
1	3.047e−2	3.043	8.329e−3	3.363	7.614e−4	4.068
2	5.638e−3	2.434	1.005e−3	3.051	7.334e−5	3.376
3	7.160e−4	2.977	1.221e−4	3.041	4.697e−6	3.965
4	8.983e−5	2.995	1.511e−5	3.014	2.954e−7	3.991
5	1.124e−5	2.999	1.882e−6	3.006	1.850e−8	3.997
6	1.405e−6	3.000	2.348e−7	3.003	1.157e−9	3.999

2.4.2 Channel flow problem in packed bed reactors

Let the reactor channel be represented by $\Omega = (0, L) \times (-R, R)$ where $R = 5$ and $L = 60$. In all computations we use the porosity distribution which is determined experimentally and takes into account the effect of wall channelling in packed bed reactors

$$\varepsilon(x,y) = \varepsilon(y) = \varepsilon_\infty \left\{ 1 + \frac{1-\varepsilon_\infty}{\varepsilon_\infty} e^{-6(R-|y|)} \right\} . \tag{2.127}$$

We distinguish between the inlet, outlet and membrane parts of domain boundary Γ, and denote them by Γ_{in}, Γ_{out} and Γ_w, respectively. Let

$$\begin{aligned} \Gamma_{in} &= \{(x,y) \in \Gamma : \; x = 0\} , \\ \Gamma_{out} &= \{(x,y) \in \Gamma : \; x = L\} , \\ \Gamma_w &= \{(x,y) \in \Gamma : \; y = -R, \; y = R\} . \end{aligned}$$

At the inlet Γ_{in} and at the membrane wall we prescribe Dirichlet boundary conditions, namely the plug flow conditions

$$\boldsymbol{u}|_{\Gamma_{in}} = \boldsymbol{u}_{in} = (u_{in}, 0)^T ,$$

and

$$\boldsymbol{u}|_{\Gamma_w} = \boldsymbol{u}_w = \begin{cases} (0, u_w)^T & \text{for} \quad y = -R , \\ (0, -u_w)^T & \text{for} \quad y = R , \end{cases}$$

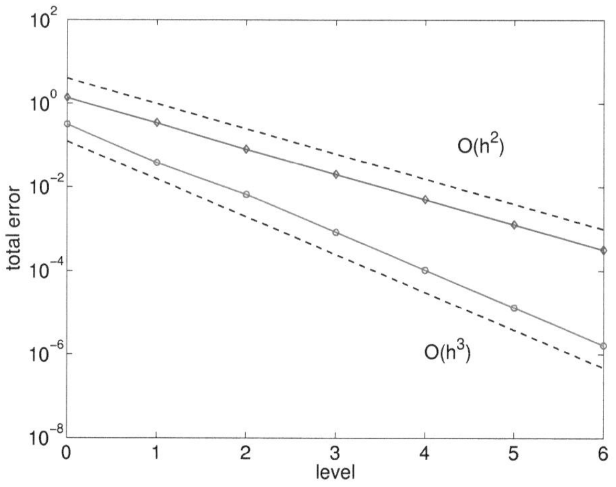

Figure 2.1: Stokes-like problem: discretisation error.

whereby $u_{in} > 0$, $u_w > 0$. At the outlet Γ_{out} we set the following outflow boundary condition

$$-\frac{1}{Re}\frac{\partial \boldsymbol{u}}{\partial \boldsymbol{n}} + p\boldsymbol{n} = \boldsymbol{0}$$

where $\boldsymbol{n}$ denotes the outer normal. In order to avoid discontinuity between the inflow and wall conditions we replace constant profile by trapezoidal one with zero value at the corners. Our computations are carried out on the cartesian mesh which on the coarse level consists of 20 stretched rectangular cells (see Figure 2.3) and will be three times uniformly refined. The plots of velocity magnitude in fixed bed reactor ($u_w = 0$) are presented along the vertical axis $x = 50$. In the investigated reactor the inlet velocity is assumed to be normalised ($u_{in} = 1$). Due to the variation of porosity we might expect higher velocity at the reactor walls Γ_w. This tunnelling effect can be well observed in Figure 2.4 which shows the velocity profiles for different Reynolds numbers. We remark that the maximum of velocity magnitude decreases with increasing Reynolds numbers. Next, we investigated the effect of dosing through the reactor wall Γ_w. This configuration corresponds to the packed bed reactor ($u_w > 0$). The uniform dosing becomes manifest in self-similar velocity profiles which are presented in Figure 2.5. The self-similarity of velocity profiles in packed bed reactor can be also confirmed by Figure 2.6 which shows the maxima of velocity magnitude

$$u_{max}(x) := \max_{-5 \le y \le 5} |u(x, y)| .$$

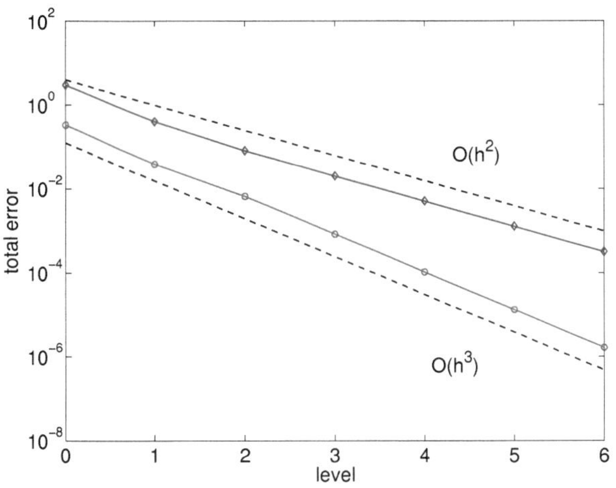

Figure 2.2: Brinkman problem: discretisation error.

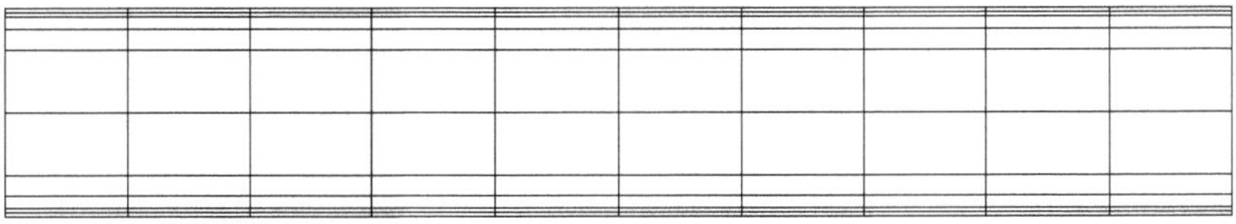

Figure 2.3: Initial mesh for reactor flow problem.

The considered maximal values of velocity grow almost linearly. The slopes given in the legend of Figure 2.6 correspond to the ratio of the wall velocity to the inlet velocity. We call the flow in fixed bed reactor developed if the vertical component of velocity vanishes and the horizontal component of velocity depends only on the vertical direction, i.e. $\boldsymbol{u}(x, y) = (u(y), 0)^T$. In this case one can show that the pressure depends linearly on the length coordinate. The short developing zone which is characteristic of this type of reactors can be recognised in Figures 2.7 and 2.8 whereas the linear distribution of the pressure is demonstrated in Figure 2.9. We refer to [76] for discussion concerning a short length of developing zone in fixed bed reactors.

The effect of dosing through membrane wall in packed bed membrane reactors can be

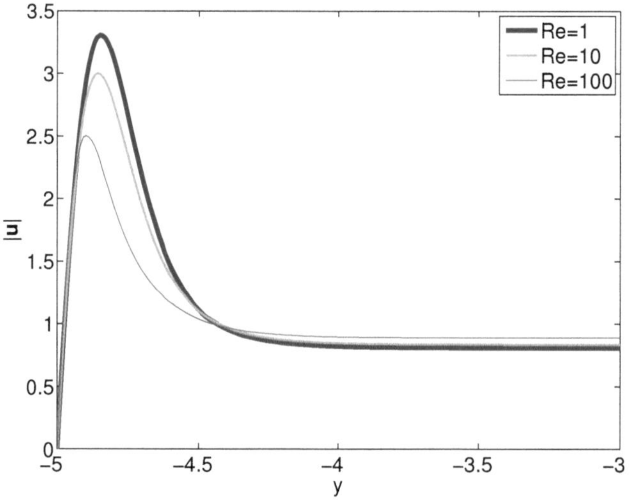

Figure 2.4: Flow profiles in fixed bed reactor at $x = 50$.

observed in Figure 2.10 which shows the magnitude of velocity. The inlet velocity is set to be $u_{in} = 0.1$ and the wall velocity u_w is chosen such that the total mass flux $(\varepsilon \boldsymbol{u}) \cdot \boldsymbol{n}$ at the position $x = 50$ is equal to that in the fixed bed reactor without dosing. The vertical component of velocity and its depth of penetration is presented in Figure 2.12. We remark that the pressure in packed bed membrane reactor seems to be superlinear, see Figure 2.11. The impact of dosing onto the pressure distribution can be observed in Figure 2.13 which shows pressure profiles along the axis $y = 0$ for different Reynolds numbers. Next, we present results for fixed bed reactors with constant porosity

$$\bar{\varepsilon} = \frac{1}{10} \int_{-5}^{5} \varepsilon(y)\, dy\,.$$

In contrast to reactors with varying porosity the velocity is almost constant with the exception of a small boundary layer near the reactor walls, see Figures 2.14 and 2.15. The linear character of pressure in fixed bed reactors is once again demonstrated in Figure 2.16. In Figure 2.17 we can see how the varying porosity affects the velocity maxima and how fast the velocity gets developed. The porosity $\varepsilon = 1$ corresponds to reactors without packing. It is well known that in this case the constant inlet velocity develops a parabolic profile.

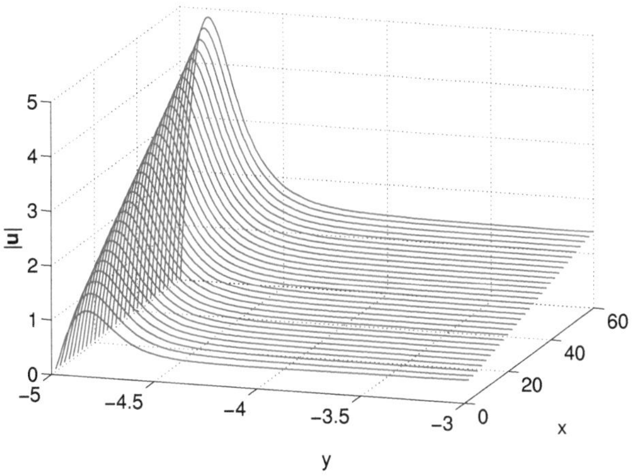

Figure 2.5: Self-similar flow profiles in packed bed membrane reactor, $Re = 10$.

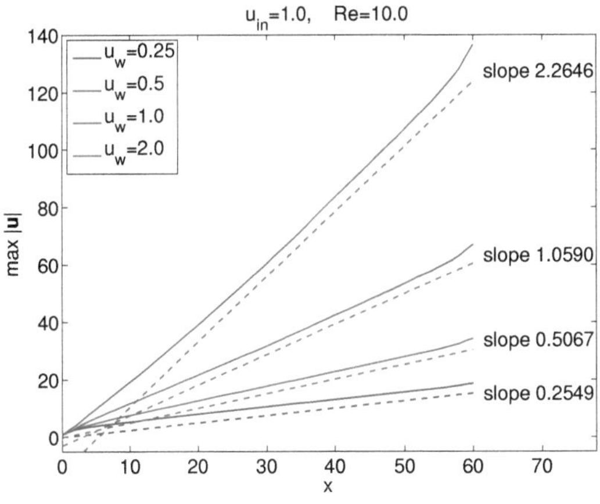

Figure 2.6: Distribution of maximum of absolute velocity along x-axis.

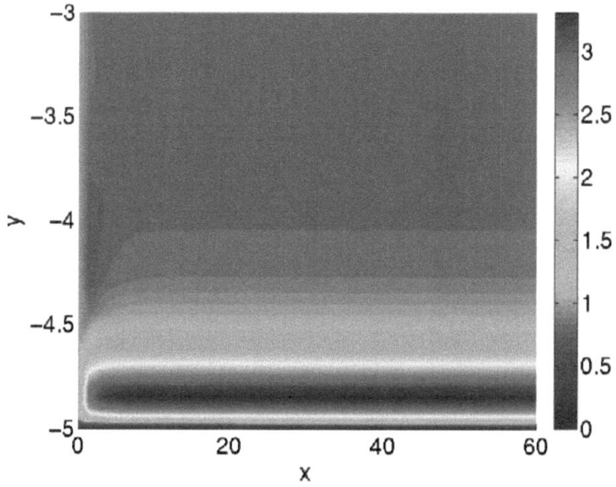

Figure 2.7: Absolute velocity in fixed bed reactor, $Re = 10$.

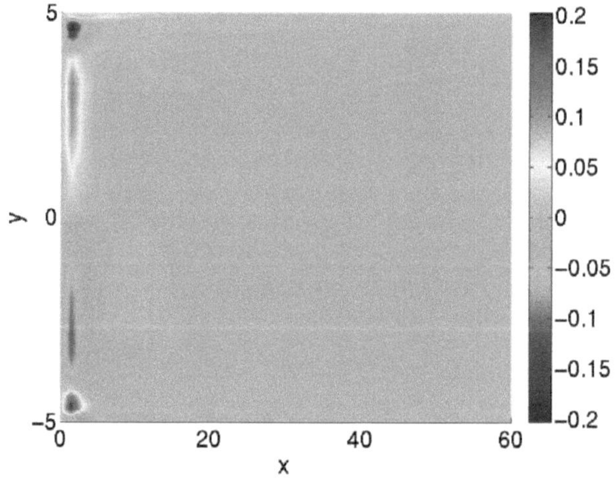

Figure 2.8: Vertical velocity in fixed bed reactor, $Re = 10$.

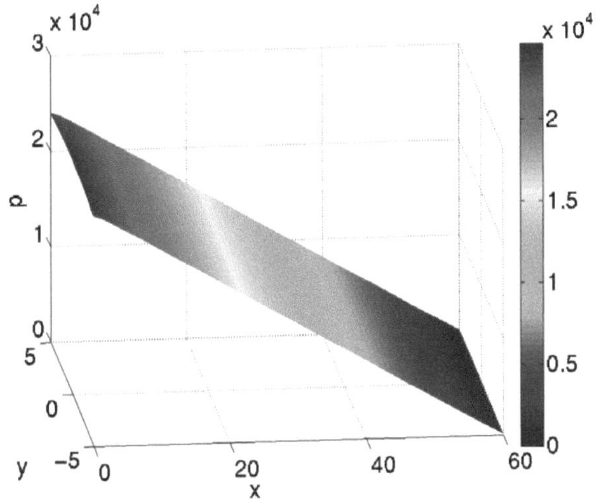

Figure 2.9: Linear pressure in fixed bed reactor, $Re = 10$.

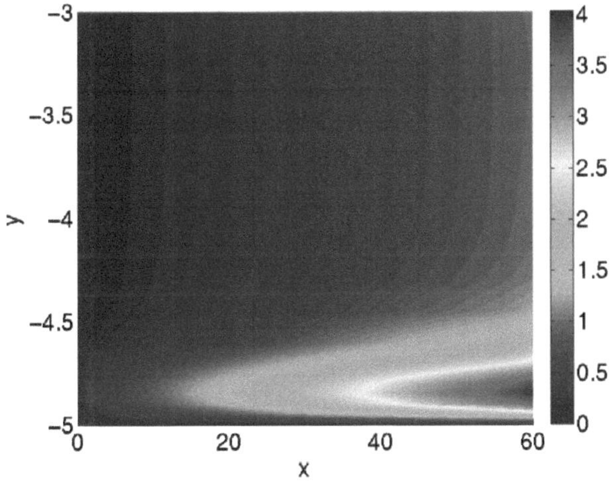

Figure 2.10: Absolute velocity in packed bed membrane reactor, $Re = 10$.

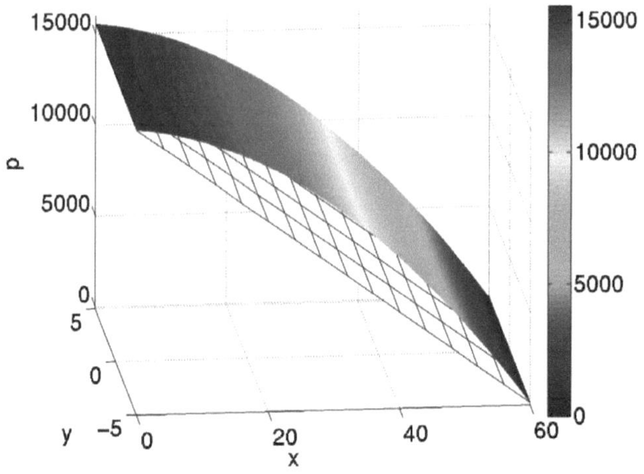

Figure 2.11: Superlinear pressure in packed bed membrane reactor, $Re = 10$.

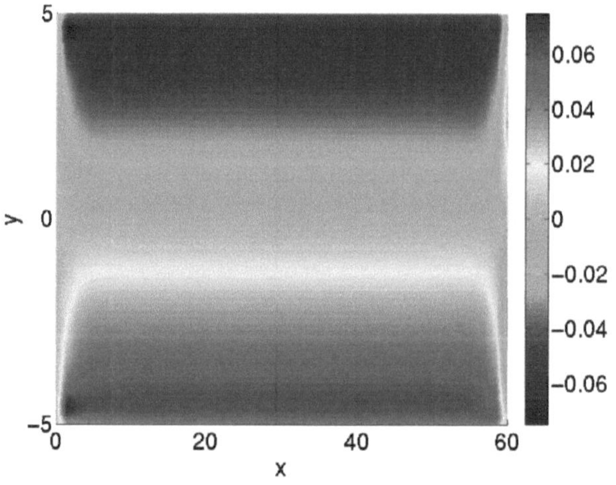

Figure 2.12: Vertical velocity in packed bed membrane reactor, $Re = 10$.

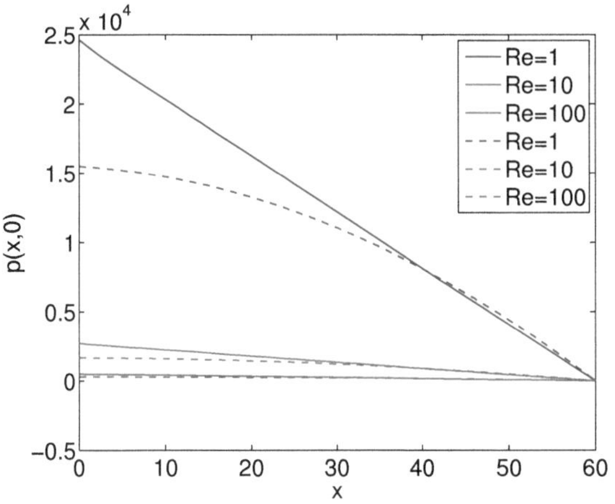

Figure 2.13: Profiles of pressure in fixed bed (solid line) and packed bed membrane reactors (dashed line) along x-axis for various Reynolds numbers.

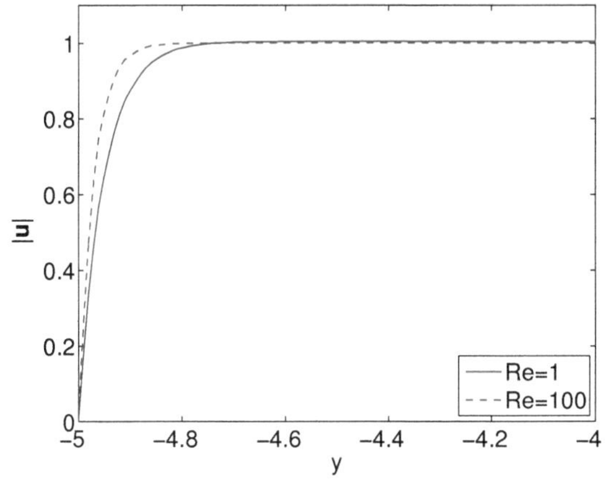

Figure 2.14: Flow profiles in fixed bed reactor with constant porosity $\bar{\varepsilon} = 0.4683$ at x=50.

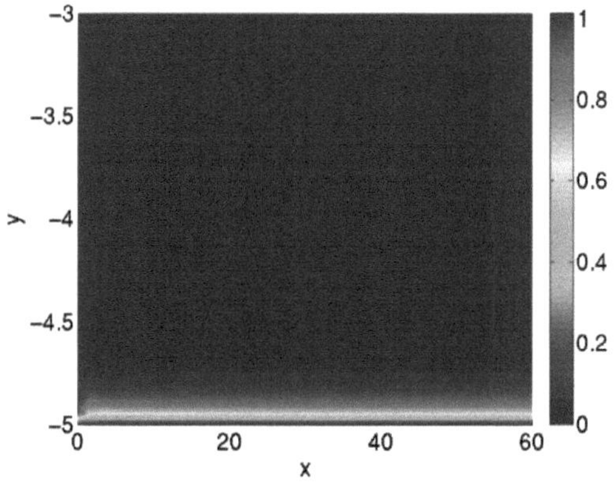

Figure 2.15: Absolute velocity in fixed bed reactor with constant porosity $\bar{\varepsilon} = 0.4683$ and $Re = 10$.

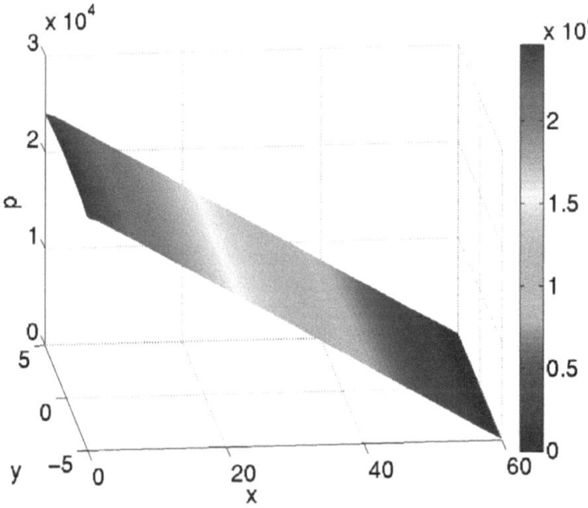

Figure 2.16: Linear pressure in fixed bed reactor with constant porosity $\bar{\varepsilon} = 0.4683$ and $Re = 1$.

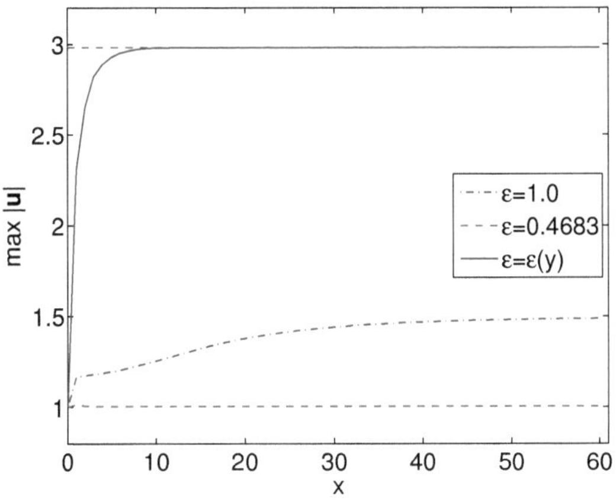

Figure 2.17: Entrance zone for $Re = 10$.

3 Stabilisation by local projection for linearised problem

3.1 Oseen-like Problem

Let us consider the linearisation of (2.5) in the form of the homogeneous Oseen–like problem

$$\left\{ \begin{array}{rcll} -\mathrm{div}\left(\frac{\varepsilon}{Re}\nabla \boldsymbol{u} - \varepsilon \boldsymbol{b}\otimes \boldsymbol{u}\right) + \varepsilon\nabla p + \left(\frac{\alpha}{Re} + \beta|\boldsymbol{b}|\right)\boldsymbol{u} &=& \boldsymbol{f} & \text{in } \Omega\,, \\ \mathrm{div}\,(\varepsilon \boldsymbol{u}) &=& 0 & \text{in } \Omega\,, \\ \boldsymbol{u} &=& \boldsymbol{0} & \text{on } \Gamma\,, \end{array} \right. \tag{3.1}$$

whereby we assume $\boldsymbol{b} \in \boldsymbol{W}^{1,\infty}(\Omega)$ and $\mathrm{div}\,(\varepsilon \boldsymbol{b}) = 0$. For the sake of abbreviation we introduce on $\boldsymbol{V}$ the bilinear form $\mathcal{A}$ given by

$$\mathcal{A}\big((\boldsymbol{u},p),(\boldsymbol{v},q)\big) := a(\boldsymbol{u},\boldsymbol{v}) + \tilde{n}(\boldsymbol{b},\boldsymbol{u},\boldsymbol{v}) + (s\boldsymbol{u},\boldsymbol{v}) - b(\boldsymbol{v},p) + b(\boldsymbol{u},q)\,.$$

Hereby, we set

$$s := \frac{1}{Re}\alpha + \beta|\boldsymbol{b}| \geq 0\,.$$

Then, the corresponding weak formulation of (3.1) reads as follows

$$\begin{array}{c} \text{Find } (\boldsymbol{u},p) \in \boldsymbol{V} \text{ such that for all } (\boldsymbol{v},q) \in \boldsymbol{V}: \\ \mathcal{A}\big((\boldsymbol{u},p);(\boldsymbol{v},q)\big) = (\boldsymbol{f},\boldsymbol{v}). \end{array} \tag{3.2}$$

We note that if $\varepsilon = 1$ then $\alpha = \beta = 0$ due to (2.6)-(2.7), and consequently $s = 0$. Then, (3.1) has the usual form of Oseen equation. The property

$$\tilde{n}(\boldsymbol{b},\boldsymbol{u},\boldsymbol{u}) = 0$$

which holds due to (2.9) allows us to apply the Lax–Milgram Lemma in the subspace $\boldsymbol{W}$ and to establish a unique velocity field $\boldsymbol{u}$. A unique pressure $p \in M$ such that $(\boldsymbol{u},p) \in \boldsymbol{V}$ solves (3.2) follows from the inf-sup condition (2.31).

3.2 Galerkin discretisation

For the finite element discretisation of the Oseen problem (3.1), we use a shape regular decomposition $\mathcal{T}_h$ of Ω into n-dimensional quadrilaterals or hexahedrons from the previous chapter. Let $Y_h \subset H^1(\Omega)$ be a finite element space of continuous, piecewise polynomial functions defined over $\mathcal{T}_h$ in Subsection 2.3.1.

Assumption A1: There exists an interpolation operator $i_h : H^1(\Omega) \to Y_h$ such that $i_h : H_0^1(\Omega) \to Y_h \cap H_0^1(\Omega)$ and

$$\|w - i_h w\|_{0,K} + h_K |w - i_h w|_{1,K} \leq C\, h_K^l \|w\|_{l,\omega(K)} \qquad \forall\, w \in H^l(\omega(K)),\ \forall\, K \in \mathcal{T}_h,\quad 1 \leq l \leq k+1\,, \tag{3.3}$$

where $\omega(K)$ denotes a certain local neighbourhood of K which appears in the definition of these interpolation operators for non-smooth functions, see [19, 70] for more details. □

We will also apply this type of interpolation operator to vector-valued functions in a component-wise manner. In this case we use boldfaced symbols like $\boldsymbol{i}_h : \boldsymbol{X}_0 \to Y_h^n \cap \boldsymbol{X}_0$. The construction of such an interpolation operator has been explained in Subsection 2.3.2, see also [3, 19, 70]. In the following, we consider the case of equal-order interpolation, thus assuming

$$\boldsymbol{X}_{h0} := Y_h^n \cap \boldsymbol{X}_0 \qquad \text{and} \qquad M_h := Y_h \cap L_0^2(\Omega)\,.$$

We set

$$\boldsymbol{V}_h := \boldsymbol{X}_{h0} \times M_h\,.$$

Now, the standard Galerkin discretisation of (3.2) reads:

$$\begin{aligned} &\text{Find } (\boldsymbol{u}_h, p_h) \in \boldsymbol{V}_h \text{ such that for all } (\boldsymbol{v}_h, q_h) \in \boldsymbol{V}_h \ : \\ &\qquad \mathcal{A}\big((\boldsymbol{u}_h, p_h); (\boldsymbol{v}_h, q_h)\big) = (\boldsymbol{f}, \boldsymbol{v}_h)\,. \end{aligned} \tag{3.4}$$

It is a well known fact that the discretisation of the Oseen-like problem by finite element methods may suffer from two reasons:

- the violation of the discrete inf-sup condition (2.48)

$$\exists \gamma > 0 : \qquad \inf_{q_h \in M_h} \sup_{\boldsymbol{v}_h \in \boldsymbol{X}_{h0}} \frac{(\operatorname{div}(\varepsilon \boldsymbol{v}_h), q_h)}{\|q_h\|_0\, |\boldsymbol{v}_h|_1} \geq \gamma \qquad \forall\, h > 0\,, \tag{3.5}$$

- the dominating advection in case of $Re \gg 1$.

In general, these two shortcomings of Galerkin discretisation lead to unphysical oscillations of the discrete solution.

3.3 Local projection stabilisation

The local projection method has been originally designed for equal-order interpolation for the Stokes problem in [5], extended to the transport equation in [6], and analysed for low order discretisations of the Oseen equations in [11]. It allows to stabilise pressure and velocity by separate terms. In [11], the case of low order Q_k-elements ($k = 1, 2$) on quadrilaterals ($n = 2$) and hexahedrons ($n = 3$) has been considered. There, the projection onto the large scale finite element space has been chosen to be the L^2-projection onto the space of discontinuous Q_{k-1}-elements on a coarser mesh. Unfortunately, this two-level approach leads to a stencil being less compact than for the SUPG/PSPG-type stabilisation. We propose to handle both instability phenomena by the local projection technique based on enrichment. It is well known that stabilised methods can also be derived from a variational multiscale formulation [39, 41, 42, 75]. Based on a scale separation of the underlying finite element spaces, it has been shown that it is sufficient to stabilise only the fine scale fluctuations. This results into a stabilising term which gives a weighted L^2-control over the gradient of fluctuations instead of the fluctuations of gradients [23, 32].

In the following, we apply results of [62] to the Oseen-like problem (3.1). Our analysis is restricted to the method based on enrichment of velocity spaces. The extension to the approach based on projection onto the coarser meshes will be not discussed in this work. Now, we want to explain the main ingredients of the local projection scheme.

Let D_h denote a discontinuous finite element space defined on the decomposition $\mathcal{T}_h$ and

$$D_h(K) := \{q_h\big|_K \; : \; q_h \in D_h\}\,.$$

Further, let

$$\pi_K : L^2(K) \to D_h(K)$$

be a local projection which defines the projection $\pi_h : L^2(\Omega) \to D_h$ by $(\pi_h w)\big|_K := \pi_K(w\big|_K)$. Associated with the projection π_h is the fluctuation operator $\kappa_h : L^2(\Omega) \to L^2(\Omega)$ defined by

$$\kappa_h := id - \pi_h\,,$$

where $id : L^2(\Omega) \to L^2(\Omega)$ is the identity. As in the previous subsection, we apply these operators to vector-valued functions in a component-wise manner and indicate this by using boldface notations, e.g. $\boldsymbol{\pi}_h : \boldsymbol{L}^2(\Omega) \to \boldsymbol{D}_h$ and $\boldsymbol{\kappa}_h : \boldsymbol{L}^2(\Omega) \to \boldsymbol{L}^2(\Omega)$.

Assumption A2: Let the fluctuation operator κ_h satisfy the following approximation property:

$$\|\kappa_h q\|_{0,K} \le C\, h_K^l |q|_{l,K} \qquad \forall\, q \in H^l(K),\ \forall\, K \in \mathcal{T}_h,\ 0 \le l \le k\,. \tag{3.6}$$

□

In the following, we define π_h as the L^2-projection in D_h. In the case of $D_h(K)$ containing the space $P_{k-1}(K)$ of polynomials of degree less than or equal to $k-1$, $k \geq 1$, we have

$$(\pi_h w - w, w_h) = 0 \qquad \forall\, w_h \in D_h,\ w \in L^2(\Omega), \tag{3.7}$$

and

$$\bigoplus_{K \in \mathcal{T}_h} P_{k-1}(K) \subset D_h. \tag{3.8}$$

Since D_h is discontinuous over the element faces, (3.7) can be localised and $\pi_K : L^2(K) \to D_h(K)$ is locally defined by

$$(\pi_K w - w, w_h)_K = 0 \qquad \forall\, w_h \in D_h(K),\ w \in L^2(K). \tag{3.9}$$

In this case, the L^2-projection $\pi_K : L^2(K) \to D_h(K)$ becomes the identity on the subspace $P_{k-1}(K) \subset H^l(K)$. Now, the Bramble–Hilbert lemma gives the approximation properties for $\kappa_h = id - \pi_h$ stated in assumption A2.

We will modify the discrete problem (3.4) by adding the stabilisation term

$$S_h\big((\boldsymbol{u}_h, p_h); (\boldsymbol{v}_h, q_h)\big) := \sum_{K \in \mathcal{T}_h} \Big\{ \tau_K \big(\boldsymbol{\kappa}_h(\nabla \boldsymbol{u}_h), \boldsymbol{\kappa}_h(\nabla \boldsymbol{v}_h)\big)_K + \alpha_K \big(\boldsymbol{\kappa}_h(\nabla p_h), \boldsymbol{\kappa}_h(\nabla q_h)\big)_K \Big\}, \tag{3.10}$$

where τ_K and α_K are user-chosen constants. Their optimal mesh-dependent choice will follow from the error analysis of the method. Now, our stabilised scheme reads:

Find $(\boldsymbol{u}_h, p_h) \in \boldsymbol{V}_h$ such that for all $(\boldsymbol{v}_h, q_h) \in \boldsymbol{V}_h$:

$$\mathcal{A}\big((\boldsymbol{u}_h, p_h); (\boldsymbol{v}_h, q_h)\big) + S_h\big((\boldsymbol{u}_h, p_h); (\boldsymbol{v}_h, q_h)\big) = (\boldsymbol{f}, \boldsymbol{v}_h). \tag{3.11}$$

Existence, uniqueness, and convergence properties of the solutions $(\boldsymbol{u}_h, p_h) \in \boldsymbol{V}_h$ will be studied in the next section.

3.4 Convergence analysis

3.4.1 Special interpolant

The key ingredient of the error analysis of the local projection method is the construction of an interpolant $j_h : H^1(\Omega) \to Y_h$ such that the error $w - j_h w$ is L^2-orthogonal to D_h without loosing the standard approximation properties. Let us define

$$Y_h(K) := \{ w_h|_K \ :\ w_h \in Y_h,\ w_h = 0 \text{ on } \Omega \backslash K \}\,.$$

Assumption A3: Let the local inf-sup condition

$$\exists\,\beta_1 > 0, \quad \forall\, h > 0 \ \forall\, K \in \mathcal{T}_h : \qquad \inf_{q_h \in D_h(K)} \sup_{v_h \in Y_h(K)} \frac{(v_h, q_h)_K}{\|v_h\|_{0,K}\, \|q_h\|_{0,K}} \geq \beta_1 > 0 \qquad (3.12)$$

be satisfied. □

We note that a necessary requirement on the spaces $Y_h(K)$ and $D_h(K)$ is

$$\dim Y_h(K) \geq \dim D_h(K). \qquad (3.13)$$

Since the spaces $Y_h(K)$ and $D_h(K)$ are defined on the same mesh, $D_h(K)$ will be chosen such that A2 holds and $Y_h(K)$ will be enriched by additional functions to fulfil A3.

Now, we recall the main theorem concerning the existence of the special interpolation operator.

Theorem 3.1 *Let assumptions A1, A3 be satisfied. Then, there are interpolation operators $j_h : H^1(\Omega) \to Y_h$ and $\boldsymbol{j}_h : \boldsymbol{X}_0 \to \boldsymbol{X}_{h0}$ satisfying the following orthogonality and approximation properties:*

$$(w - j_h w, q_h) = 0 \qquad \forall\, q_h \in D_h,\ \forall\, w \in H^1(\Omega)\,, \qquad (3.14)$$

$$\|w - j_h w\|_{0,K} + h_K |w - j_h w|_{1,K} \leq C\, h_K^l \|w\|_{l,\omega(K)} \qquad \forall\, w \in H^l(\Omega)\,,\ 1 \leq l \leq k+1\,, \quad \forall\, K \in \mathcal{T}_h\,, \qquad (3.15)$$

$$(\boldsymbol{w} - \boldsymbol{j}_h \boldsymbol{w}, \boldsymbol{q}_h) = 0 \qquad \forall\, \boldsymbol{q}_h \in \boldsymbol{D}_h,\ \forall\, \boldsymbol{w} \in \boldsymbol{V}\,, \qquad (3.16)$$

$$\|\boldsymbol{w} - \boldsymbol{j}_h \boldsymbol{w}\|_{0,K} + h_K |\boldsymbol{w} - \boldsymbol{j}_h \boldsymbol{w}|_{1,K} \leq C\, h_K^l \|\boldsymbol{w}\|_{l,\omega(K)} \qquad \forall\, \boldsymbol{w} \in \boldsymbol{X}_0 \cap \boldsymbol{H}^l(\Omega)\,, \quad 1 \leq l \leq k+1,\ \forall\, K \in \mathcal{T}_h\,. \qquad (3.17)$$

Proof. See the proof of Theorem 2.2 in [62]. □

Projection spaces based on mapped finite elements

Let us introduce

$$P_{k-1,h}^{\mathrm{disc}} := \{ v \in L^2(\Omega) \ :\ v|_K \circ \boldsymbol{F}_K \in P_{k-1}(\widehat{K}) \quad \forall\, K \in \mathcal{T}_h \}$$

finite element spaces for the projection space D_h. In order to obtain the optimal order of the interpolation error for the mapped projection space, families of uniformly refined quadrilateral/hexahedral meshes are required, see again [4, 60]. We extend the approximation spaces in order to ensure the local inf-sup condition A3. To this end, let

$$\hat{b}(\hat{\boldsymbol{x}}) = \prod_{i=1}^{n}(1-\hat{\boldsymbol{x}}_i^2) \in Q_2(\widehat{K}), \quad \hat{\boldsymbol{x}} = (\hat{x}_1,\ldots,\hat{x}_n) \in \widehat{K}, \quad n = 2,3, \tag{3.18}$$

be a bubble function associated with the reference cell $\widehat{K} = (-1,1)^n$. The enriched finite element space is set to be

$$Q_k^+(\widehat{K}) := Q_k(\widehat{K}) \oplus \mathrm{span}\left\{\hat{b}\,\hat{x}_i^{k-1}, \quad i = 1,\ldots,n\right\}.$$

We define a pair of finite element spaces

$$(Y_h, D_h) := (Q_{k,h}^+, P_{k-1,h}^{\mathrm{disc}})$$

via the reference mapping

$$Q_{k,h}^+ := \{v \in H^1(\Omega) \,:\, v|_K \circ \boldsymbol{F}_K \in Q_k^+(\widehat{K}) \qquad \forall\, K \in \mathcal{T}_h\}.$$

We note that in general the functions of spaces $Q_{k,h}^+$, $P_{k-1,h}^{\mathrm{disc}}$ are not polynomials. Since $Q_k(\widehat{K}) \subset Q_k^+(\widehat{K})$, the assumption A1 is satisfied. Assumption A2 holds on uniformly refined meshes, see [4, 60].

Lemma 3.2 *Let the local projection scheme be defined for the pair* $(Y_h, D_h) = (Q_{k,h}^+, P_{k-1}^{disc})$ *with an arbitrary but fixed polynomial degree* $k \in \mathbb{N}$. *Then, the local inf-sup condition A3 holds with a constant* β_1 *independent of* h.

Proof. For an arbitrary $q \in D_h(K)$ we choose $v(\boldsymbol{x}) := (\hat{q}\cdot\hat{b}) \circ \boldsymbol{F}_K^{-1}(\boldsymbol{x})$ where $\hat{b} \geq 0$ is the nonnegative bubble function from (3.18), $\hat{q} \in P_{k-1}(\widehat{K})$. Since $\hat{q} = \hat{q}_0 + \hat{q}_1$ with $\hat{q}_0 \in \mathrm{span}\left\{x_i^{k-1}, \quad i = 1,\ldots,n\right\}$ and $\hat{q}_1 \in Q_{k-2}$, we have $\hat{v}(\hat{\boldsymbol{x}}) := \hat{q}(\hat{\boldsymbol{x}})\hat{b}(\hat{\boldsymbol{x}}) \in Q_k^+(\widehat{K})$. Moreover, we have $\hat{v}(\hat{\boldsymbol{x}}) \in \widehat{Y}(\widehat{K})$, since $\hat{b}|_{\partial\widehat{K}} = 0$. Then, it follows from the estimate (2.44)

$$\begin{aligned}
(q,v)_K &= \int_K q(\boldsymbol{x})v(\boldsymbol{x})\,d\boldsymbol{x} = \int_{\widehat{K}} \hat{q}(\hat{\boldsymbol{x}})\hat{v}(\hat{\boldsymbol{x}})\,|\det D\boldsymbol{F}_K(\hat{\boldsymbol{x}})|\,d\hat{\boldsymbol{x}} \\
&= \int_{\widehat{K}} \hat{q}(\hat{\boldsymbol{x}})\hat{q}(\hat{\boldsymbol{x}})\hat{b}(\hat{\boldsymbol{x}})\,|\det D\boldsymbol{F}_K(\hat{\boldsymbol{x}})|\,d\hat{\boldsymbol{x}} \\
&\geq Cn!(1-\gamma_K)^n\, h_K^n \int_{\widehat{K}} (\hat{q}(\hat{\boldsymbol{x}}))^2\,\hat{b}(\hat{\boldsymbol{x}})\,d\hat{\boldsymbol{x}}\,.
\end{aligned}$$

The equivalence of norms on the finite dimensional space $Q_{k-1}(\widehat{K})$ implies

$$\|\hat{q}\cdot\sqrt{\hat{b}}\|_{0,\widehat{K}} \geq C\|\hat{q}\|_{0,\widehat{K}} \qquad \forall\,\hat{q} \in Q_{k-1}(\widehat{K})$$

and hence

$$(q,v)_K \geq Cn!(1-\gamma_K)^n\, h_K^n\, \|\hat q\|^2_{0,\widehat K}\,. \tag{3.19}$$

Using $|\hat b(\hat{\boldsymbol x})| \leq 1 \quad \forall\, \hat{\boldsymbol x} \in \widehat K$, we get

$$\|v\|^2_{0,K} \leq \int_{\widehat K} (\hat q(\hat{\boldsymbol x}))^2\, |\det D\boldsymbol F_M(\hat{\boldsymbol x})|\, d\hat{\boldsymbol x} \leq Cn!(1+\gamma_K)^n\, h_K^n \|\hat q\|^2_{0,\widehat K}\,. \tag{3.20}$$

Evoking (2.43), we obtain

$$\|q\|^2_{0,K} \leq Cn!(1+\gamma_K)^n h_K^n \|\hat q\|^2_{0,\widehat K} \qquad \forall q \in D_h(K)\,. \tag{3.21}$$

From (3.19)–(3.21) it follows immediately

$$\forall\, q \in D_h(K) \quad \exists\, v \in Y_h(K): \qquad \frac{(q,v)_K}{\|q\|_{0,K}\|v\|_{0,K}} \geq C\left(\frac{1-\gamma_K}{1+\gamma_K}\right)^n \geq C\left(\frac{1-\gamma}{1+\gamma}\right)^n =: \beta_1\,.$$

This implies the local inf-sup condition A3. □

Remark 3.3 *A comparison of the dimensions of the spaces $Y_h(K)$ and $D_h(K)$ shows that*

$$\dim \widehat Y(\widehat K) = (k-1)^n + n \geq \binom{k-1+n}{n} = \dim P_{k-1}(\widehat K) \qquad \forall\, k \in \mathbb N \quad \forall\, n \in \mathbb N\,.$$

In particular, the enrichment is optimal for biquadratic and bicubic elements on quadrilaterals and for triquadratic elements on hexahedra.

Remark 3.4 *Note that the space $Q_k^+(\widehat K)$ has for $k \geq 2$ exactly n basis functions more than $Q_k(\widehat K)$, independent of k.*

3.4.2 Stability

Let us introduce the mesh-dependent norm on the product space $\boldsymbol V$ by

$$|||(\boldsymbol v,q)||| := \Big(Re^{-1}\|\sqrt{\varepsilon}\nabla\boldsymbol v\|_0^2 + \|\sqrt{s}\boldsymbol v\|_0^2 + (Re^{-1} + \|s\|_{0,\infty})\|q\|_0^2 + S_h\big((\boldsymbol v,q);(\boldsymbol v,q)\big)\Big)^{1/2}. \tag{3.22}$$

We show that the bilinear form $(\mathcal A + S_h)$ satisfies an inf-sup condition on $\boldsymbol V_h$.

Lemma 3.5 *Assume A1, A3, $\varepsilon \in W^{1,\infty}(\Omega)$ and $\max(Re^{-1}, \|s\|_{0,\infty}, \tau_K, h_K^2/\alpha_K) \leq C$ for all $K \in \mathcal T_h$. Then, there is a positive constant β_2 independent of Re^{-1} and h such that*

$$\inf_{(\boldsymbol v_h,q_h)\in \boldsymbol V_{h0}} \sup_{(\boldsymbol w_h,r_h)\in \boldsymbol V_{h0}} \frac{(\mathcal A + S_h)\big((\boldsymbol v_h,q_h);(\boldsymbol w_h,r_h)\big)}{|||(\boldsymbol v_h,q_h)|||\;|||(\boldsymbol w_h,r_h)|||} \geq \beta_2 > 0 \tag{3.23}$$

holds true.

Proof. Let us consider an arbitrary $(\boldsymbol{v}_h, q_h) \in \boldsymbol{V}_{h0}$. Choosing $(\boldsymbol{w}_h, r_h) = (\boldsymbol{v}_h, q_h)$, we have

$$(\mathcal{A} + S_h)\big((\boldsymbol{v}_h, q_h); (\boldsymbol{v}_h, q_h)\big) = Re^{-1}\|\sqrt{\varepsilon}\nabla \boldsymbol{v}_h\|_0^2 + \|\sqrt{s}\boldsymbol{v}_h\|_0^2 + S_h\big((\boldsymbol{v}_h, q_h); (\boldsymbol{v}_h, q_h)\big) \quad (3.24)$$

due to property (2.9).

Now we consider another choice to generate an L^2-norm control over the pressure. For any $q_h \in M_h$, the continuous inf-sup condition (2.31) guarantees the existence of a function $\boldsymbol{v}_{q_h} \in \boldsymbol{X}_0$ such that

$$\big(\operatorname{div}(\varepsilon \boldsymbol{v}_{q_h}), q_h\big) = -(q_h, q_h), \qquad \|\boldsymbol{v}_{q_h}\|_1 \le C\|q_h\|_0 . \quad (3.25)$$

We choose $(\boldsymbol{w}_h, r_h) = (\boldsymbol{j}_h \boldsymbol{v}_{q_h}, 0)$ where $\boldsymbol{j}_h$ is the interpolant of Theorem 3.1 satisfying (3.16) and (3.17). Thus, we obtain

$$\begin{aligned}\mathcal{A}\big((\boldsymbol{v}_h, q_h); (\boldsymbol{j}_h \boldsymbol{v}_{q_h}, 0)\big) = \|q_h\|_0^2 &- \big(q_h, \operatorname{div}(\varepsilon(\boldsymbol{j}_h \boldsymbol{v}_{q_h} - \boldsymbol{v}_{q_h}))\big) + \big((\varepsilon \boldsymbol{b}\cdot\nabla)\boldsymbol{v}_h, \boldsymbol{j}_h \boldsymbol{v}_{q_h}\big) \\ &+ Re^{-1}(\varepsilon\nabla \boldsymbol{v}_h, \nabla \boldsymbol{j}_h \boldsymbol{v}_{q_h}) + (s\boldsymbol{v}_h, \boldsymbol{j}_h \boldsymbol{v}_{q_h}) .\end{aligned} \quad (3.26)$$

We estimate the last four terms on the right hand side. Starting with an integration by parts of the first term, we get

$$-\big(q_h, \operatorname{div}(\varepsilon(\boldsymbol{j}_h \boldsymbol{v}_{q_h} - \boldsymbol{v}_{q_h}))\big) = \big(\varepsilon\nabla q_h, (\boldsymbol{j}_h \boldsymbol{v}_{q_h} - \boldsymbol{v}_{q_h})\big) = \big(\boldsymbol{\kappa}_h(\varepsilon\nabla q_h), (\boldsymbol{j}_h \boldsymbol{v}_{q_h} - \boldsymbol{v}_{q_h})\big) . \quad (3.27)$$

Now, let $\bar{\varepsilon}$ be the L^2-projection of ε in the space of piecewise constant functions with respect to the decomposition $\mathcal{T}_h$. Using the L^2-stability of $\boldsymbol{\kappa}_h$, Bramble–Hilbert lemma, an inverse inequality, $\boldsymbol{\kappa}_h(\bar{\varepsilon}\nabla q_h) = \bar{\varepsilon}\boldsymbol{\kappa}_h(\nabla q_h)$, we get for $\varepsilon \in W^{1,\infty}(\Omega)$

$$\begin{aligned}\big\|\boldsymbol{\kappa}_h(\varepsilon\nabla q_h)\big\|_{0,K} &\le \big\|\boldsymbol{\kappa}_h((\varepsilon - \bar{\varepsilon})\nabla q_h)\big\|_{0,K} + \big\|\boldsymbol{\kappa}_h(\bar{\varepsilon}\nabla q_h\big\|_{0,K} \\ &\le C\, h_K |\varepsilon|_{1,\infty,K}\|\nabla q_h\|_{0,K} + \|\varepsilon\|_{0,\infty,K}\big\|\boldsymbol{\kappa}_h(\nabla q_h)\big\|_{0,K} \\ &\le C|\varepsilon|_{1,\infty,K}\|q_h\|_{0,K} + \big\|\boldsymbol{\kappa}_h(\nabla q_h)\big\|_{0,K}\end{aligned} \quad (3.28)$$

due to the model assumption (A1). Then, from (3.27) and (3.17) we deduce

$$\begin{aligned}\big|\big(q_h, \operatorname{div}(\varepsilon(\boldsymbol{j}_h \boldsymbol{v}_{q_h} - \boldsymbol{v}_{q_h}))\big)\big| &\le \Big(\sum_{K\in\mathcal{T}_h} \alpha_K\|\boldsymbol{\kappa}_h\nabla q_h\|_{0,K}^2\Big)^{1/2}\Big(\sum_{K\in\mathcal{T}_h}\frac{1}{\alpha_K}\|\boldsymbol{j}_h \boldsymbol{v}_{q_h} - \boldsymbol{v}_{q_h}\|_{0,K}^2\Big)^{1/2} \\ &\quad + C|\varepsilon|_{1,\infty}\|q_h\|_0\Big(\sum_{K\in\mathcal{T}_h}\|\boldsymbol{j}_h \boldsymbol{v}_{q_h} - \boldsymbol{v}_{q_h}\|_{0,K}^2\Big)^{1/2} \\ &\le C\,\big(S_h\big((\boldsymbol{v}_h, q_h); (\boldsymbol{v}_h, q_h)\big)\big)^{1/2}\|\boldsymbol{v}_{q_h}\|_1 + Ch|\varepsilon|_{1,\infty}\|q_h\|_0\|\boldsymbol{v}_{q_h}\|_1 \\ &\le C\,\big(S_h\big((\boldsymbol{v}_h, q_h); (\boldsymbol{v}_h, q_h)\big)\big)^{1/2}\|q_h\|_0 + Ch|\varepsilon|_{1,\infty}\|q_h\|_0^2 \\ &\le \Big(\frac{1}{16} + Ch|\varepsilon|_{1,\infty}\Big)\|q\|_0^2 + C\,S_h\big((\boldsymbol{v}_h, q_h); (\boldsymbol{v}_h, q_h)\big) \\ &\le \frac{1}{8}\|q\|_0^2 + C\,S_h\big((\boldsymbol{v}_h, q_h); (\boldsymbol{v}_h, q_h)\big) ,\end{aligned} \quad (3.29)$$

provided that $C|\varepsilon|_{1,\infty}h \leq 1/8$ holds. Integrating by parts, using the H^1 stability of j_h which follows from Theorem 3.1, and (3.25), we obtain for the third term in (3.26)

$$\begin{aligned} |((\varepsilon \boldsymbol{b} \cdot \nabla)\boldsymbol{v}_h, \boldsymbol{j}_h \boldsymbol{v}_{q_h})| &= |(\boldsymbol{v}_h, (\varepsilon \boldsymbol{b} \cdot \nabla)\boldsymbol{j}_h \boldsymbol{v}_{q_h})| \leq C\, \|\boldsymbol{v}_h\|_0\, |\boldsymbol{j}_h \boldsymbol{v}_{q_h}|_1 \\ &\leq \frac{\|q_h\|_0^2}{8} + C\, \|\boldsymbol{v}_h\|_0^2 . \end{aligned} \tag{3.30}$$

For estimating the remaining terms in (3.26), we use $\max(Re^{-1}, \|s\|_{0,\infty}) \leq C$ to get

$$\begin{aligned} \left| Re^{-1}(\varepsilon \nabla \boldsymbol{v}_h, \nabla \boldsymbol{j}_h \boldsymbol{v}_{q_h}) + (s\boldsymbol{v}_h, \boldsymbol{j}_h \boldsymbol{v}_{q_h}) \right| &\leq \left(Re^{-1} \|\sqrt{\varepsilon} \nabla \boldsymbol{v}_h\|_0 + \|\sqrt{s}\boldsymbol{v}_h\|_0 \right) \|\boldsymbol{j}_h \boldsymbol{v}_{q_h}\|_1 \\ &\leq C\left\{ Re^{-1/2} \|\sqrt{\varepsilon}\nabla \boldsymbol{v}_h\|_0 + \|\sqrt{s}\boldsymbol{v}_h\|_0 \right\} \|q_h\|_0 \\ &\leq \frac{\|q_h\|_0^2}{8} + C\left\{ Re^{-1} \|\sqrt{\varepsilon}\nabla \boldsymbol{v}_h\|_0^2 + \|\sqrt{s}\boldsymbol{v}_h\|_0^2 \right\} . \end{aligned} \tag{3.31}$$

The Cauchy–Schwarz inequality and the L^2-stability of $\boldsymbol{\kappa}_h$ give

$$\begin{aligned} \left| S_h\big((\boldsymbol{v}_h, q_h); (\boldsymbol{j}_h \boldsymbol{v}_{q_h}, 0)\big) \right| &\leq C\, \Big(S_h\big((\boldsymbol{v}_h, 0); (\boldsymbol{v}_h, 0)\big) \Big)^{1/2} |\boldsymbol{j}_h \boldsymbol{v}_{q_h}|_1 \\ &\leq C\, \Big(S_h\big((\boldsymbol{v}_h, q_h); (\boldsymbol{v}_h, q_h)\big) \Big)^{1/2} \|q_h\|_0 \\ &\leq \frac{\|q_h\|_0^2}{8} + C\, S_h\big((\boldsymbol{v}_h, q_h); (\boldsymbol{v}_h, q_h)\big). \end{aligned} \tag{3.32}$$

Let

$$X := \Big(Re^{-1} \|\sqrt{\varepsilon}\nabla \boldsymbol{v}_h\|_0^2 + \|\sqrt{s}\boldsymbol{v}_h\|_0^2 + S_h\big((\boldsymbol{v}_h, q_h); (\boldsymbol{v}_h, q_h)\big) \Big)^{1/2}$$

denote the part of the triple norm without L^2-control over the pressure. Using (3.29)–(3.32), we get from (3.26)

$$(\mathcal{A} + S_h)\big((\boldsymbol{v}_h, q_h); (\boldsymbol{j}_h \boldsymbol{v}_{q_h}, 0)\big) \geq \frac{1}{2}\|q_h\|_0^2 - C\, X^2 - C\, \|\boldsymbol{v}_h\|_0^2 \tag{3.33}$$

Now, we multiply (3.33) by $t := 2(Re^{-1} + \|s\|_{0,\infty})$ and use the Poincaré inequality and properties of ε and s to estimate

$$t\|\boldsymbol{v}_h\|_0^2 \leq C\big(Re^{-1} \|\sqrt{\varepsilon}\nabla \boldsymbol{v}_h\|_0^2 + \|\sqrt{s}\boldsymbol{v}_h\|_0^2 \big).$$

Hence, we obtain

$$(\mathcal{A} + S_h)\big((\boldsymbol{v}_h, q_h); t(\boldsymbol{j}_h \boldsymbol{v}_{q_h}, 0)\big) \geq \big(Re^{-1} + \|s\|_{0,\infty} \big) \|q_h\|_0^2 - C_1\, X^2 \tag{3.34}$$

with a suitable constant C_1. We define for an arbitrary $(\boldsymbol{v}_h, q_h) \in \boldsymbol{V}_h$

$$(\boldsymbol{w}_h, r_h) := (\boldsymbol{v}_h, q_h) + \frac{t}{1 + C_1} (\boldsymbol{j}_h \boldsymbol{v}_{q_h}, 0) \in \boldsymbol{V}_h.$$

Then, we have

$$\begin{aligned}(\mathcal{A}+S_h)\big((\boldsymbol{v}_h,q_h);(\boldsymbol{w}_h,r_h)\big) &\geq \frac{(Re^{-1}+\|s\|_{0,\infty})}{1+C_1}\|q_h\|_0^2+\left(1-\frac{C_1}{1+C_1}\right)X^2 \\ &= \frac{1}{1+C_1}|||(\boldsymbol{v}_h,q_h)|||^2\end{aligned} \tag{3.35}$$

and

$$\begin{aligned}|||(\boldsymbol{w}_h,r_h)||| &\leq |||(\boldsymbol{v}_h,q_h)|||+\frac{t}{1+C_1}|||(\boldsymbol{j}_h\boldsymbol{v}_{q_h},0)||| \\ &\leq |||(\boldsymbol{v}_h,q_h)|||+C\{Re^{-1}+\|s\|_{0,\infty}\}\|\boldsymbol{j}_h\boldsymbol{v}_{q_h}\|_1 \\ &\leq |||(\boldsymbol{v}_h,q_h)|||+C\{Re^{-1}+\|s\|_{0,\infty}\}\|q_h\|_0 \\ &\leq C_2|||(\boldsymbol{v}_h,q_h)|||.\end{aligned} \tag{3.36}$$

From (3.35) and (3.36) we state (3.23) with the inf-sup constant $\beta_2 = 1/(C_2(1+C_1))$. □

Remark 3.6 *The unique solvability of the stabilised discrete problem* (3.11) *follows directly from Lemma* 3.5.

3.4.3 Approximated Galerkin orthogonality

In contrast to residual-based stabilisation schemes [12], we do not have the Galerkin orthogonality. Therefore, we estimate the consistency error.

Lemma 3.7 *Let* $(\boldsymbol{u},p)\in\boldsymbol{V}$ *be the solution of* (3.2) *and* $(\boldsymbol{u}_h,p_h)\in\boldsymbol{V}_{h0}$ *be the solution of* (3.11)*, respectively. Then,*

$$\mathcal{A}((\boldsymbol{u}-\boldsymbol{u}_h,p-p_h);(\boldsymbol{v}_h,q_h)) = S_h((\boldsymbol{u}_h,p_h);(\boldsymbol{v}_h,q_h)) \qquad \forall(\boldsymbol{v}_h,q_h)\in\boldsymbol{V}_{h0}\,. \tag{3.37}$$

Proof. We get (3.37) simply by subtracting (3.11) from (3.2). □

Next, we estimate the consistency error.

Lemma 3.8 *Let the fluctuation operator* κ_h *satisfy A2. Then, for* $(\boldsymbol{u},p)\in\boldsymbol{H}^{k+1}(\Omega)\times H^{k+1}(\Omega)$ *we have*

$$\big|S_h\big((\boldsymbol{u},p);(\boldsymbol{v}_h,q_h)\big)\big| \leq C\left(\sum_{K\in\mathcal{T}_h} h_K^{2k}\Big[\big(\tau_K\|\boldsymbol{u}\|_{k+1,K}^2+\alpha_K\|p\|_{k+1,K}^2\big)\Big]\right)^{1/2} |||(\boldsymbol{v}_h,q_h)||| \tag{3.38}$$

for all $(\boldsymbol{v}_h,q_h)\in\boldsymbol{V}_{h0}$.

Proof. From the definition of the stabilising term we get

$$\begin{aligned}|S_h((\boldsymbol{u},p);(\boldsymbol{v}_h,q_h))| &\le \Big(S_h((\boldsymbol{u},p);(\boldsymbol{u},p))\Big)^{1/2}\Big(S_h((\boldsymbol{v}_h,q_h);(\boldsymbol{v}_h,q_h))\Big)^{1/2}\\ &\le \Big(S_h((\boldsymbol{u},p);(\boldsymbol{u},p))\Big)^{1/2}|||(\boldsymbol{v}_h,q_h)|||.\end{aligned}$$

Using the approximation properties of $\boldsymbol{\kappa}_h$, we see that

$$S_h\big((\boldsymbol{u},p);(\boldsymbol{u},p)\big) \le C\sum_{K\in\mathcal{T}_h} h_K^{2k}\big(\tau_K|\nabla\boldsymbol{u}|_{k,K}^2+\alpha_K|\nabla p|_{k,K}^2\big)$$

and (3.38) follows. □

3.4.4 A-priori error estimate

We get from stability and consistency an a-priori error estimate in the usual way. We derive error bounds with constants, which will be independent of the Reynolds number Re and h.

Theorem 3.9 *Assume A1–A3. Let* $(\boldsymbol{u},p)\in\big(\boldsymbol{H}_0^1(\Omega)\cap\boldsymbol{H}^{k+1}(\Omega)\big)\times\big(L_0^2(\Omega)\cap H^{k+1}(\Omega)\big)$ *be the weak solution of* (3.2) *and* $(\boldsymbol{u}_h,p_h)\in\boldsymbol{V}_{h0}$ *be the solution of the local projection method* (3.11). *Then, for* $s>0$ *there is a positive constant* C *independent of* Re *such that*

$$\begin{aligned}&|||(\boldsymbol{u}-\boldsymbol{u}_h,p-p_h)|||\\ &\le C\Bigg(\sum_{K\in\mathcal{T}_h} h_K^{2k}\Big[Re^{-1}+h_K^2\,\|s\|_{0,\infty,K}+h_K^2\|s\|_{0,\infty,K}^{-1}\{|\boldsymbol{b}|_{1,\infty,K}^2+|\varepsilon|_{1,\infty,K}^2\|\boldsymbol{b}\|_{0,\infty,K}^2+1\}\\ &\qquad+\tau_K^{-1}h_K^2\|\boldsymbol{b}\|_{0,\infty,K}^2+h_K^2\,\alpha_K^{-1}+\tau_K+\alpha_K\Big]\Big(\|\boldsymbol{u}\|_{k+1,\omega(K)}^2+\|p\|_{k+1,\omega(K)}^2\Big)\Bigg)^{1/2}\end{aligned}\tag{3.39}$$

holds true for sufficiently small $h>0$. *The choice*

$$\tau_K\sim h_K \qquad \text{and} \qquad \alpha_K\sim h_K$$

is asymptotically optimal and leads to

$$|||(\boldsymbol{u}-\boldsymbol{u}_h,p-p_h)|||\le C_s\Bigg(\sum_{K\in\mathcal{T}_h}(Re^{-1}+h_K)\,h_K^{2k}\Big(\|\boldsymbol{u}\|_{k+1,\omega(K)}^2+\|p\|_{k+1,\omega(K)}^2\Big)\Bigg)^{1/2}\tag{3.40}$$

with a constant C_s *independent of* Re *but depending on* s.

Proof. Starting with Lemma 3.5, we get for sufficiently small $h > 0$ an estimate for the error to the interpolants:

$$\begin{aligned}
&|||(\boldsymbol{j}_h\boldsymbol{u} - \boldsymbol{u}_h, j_h p - p_h)||| \\
&\leq \frac{1}{\beta_2} \sup_{(\boldsymbol{w}_h, r_h)\in \boldsymbol{V}_{h0}} \frac{(\mathcal{A}+S_h)\big((\boldsymbol{j}_h\boldsymbol{u} - \boldsymbol{u}_h, j_h p - p_h);(\boldsymbol{w}_h, r_h)\big)}{|||(\boldsymbol{w}_h, r_h)|||} \\
&\leq \frac{1}{\beta_2} \sup_{(\boldsymbol{w}_h, r_h)\in \boldsymbol{V}_{h0}} \frac{(\mathcal{A}+S_h)\big((\boldsymbol{u} - \boldsymbol{u}_h, p - p_h);(\boldsymbol{w}_h, r_h)\big)}{|||(\boldsymbol{w}_h, r_h)|||} \\
&\quad + \frac{1}{\beta_2} \sup_{(\boldsymbol{w}_h, r_h)\in \boldsymbol{V}_{h0}} \frac{(\mathcal{A}+S_h)\big((\boldsymbol{j}_h\boldsymbol{u} - \boldsymbol{u}, j_h p - p);(\boldsymbol{w}_h, r_h)\big)}{|||(\boldsymbol{w}_h, r_h)|||}.
\end{aligned} \tag{3.41}$$

Using Lemmata 3.7 and 3.7, we estimate the first term by

$$\begin{aligned}
&(\mathcal{A}+S_h)\big((\boldsymbol{u} - \boldsymbol{u}_h, p - p_h);(\boldsymbol{w}_h, r_h)\big) = S_h\big((\boldsymbol{u}, p);(\boldsymbol{w}_h, r_h)\big) \\
&\leq C\Bigg(\sum_{K\in\mathcal{T}_h} h_K^{2k}\Big[\tau_K \|\boldsymbol{u}\|_{k+1,K}^2 + \alpha_K \|p\|_{k+1,K}^2\Big]\Bigg)^{1/2} |||(\boldsymbol{w}_h, r_h)|||.
\end{aligned} \tag{3.42}$$

For the estimation of the second term, we consider each individual term in

$$(\mathcal{A}+S_h)((\boldsymbol{j}_h\boldsymbol{u} - \boldsymbol{u}, j_h p - p);(\boldsymbol{w}_h, r_h))$$

separately. The estimation of

$$\begin{aligned}
&Re^{-1}\big(\varepsilon\nabla(\boldsymbol{j}_h\boldsymbol{u} - \boldsymbol{u}), \nabla\boldsymbol{w}_h\big) + \big(s(\boldsymbol{j}_h\boldsymbol{u} - \boldsymbol{u}), \boldsymbol{w}_h\big) \\
&\leq C\Bigg(\sum_{K\in\mathcal{T}_h} h_K^{2k}\big(Re^{-1} + \|s\|_{0,\infty,K}\, h_K^2\big)\|\boldsymbol{u}\|_{k+1,\omega(K)}^2\Bigg)^{1/2} |||(\boldsymbol{w}_h, r_h)|||
\end{aligned} \tag{3.43}$$

is standard. When estimating the next three terms, we use the interpolant constructed in Theorem 3.1. Integrating by parts, we get

$$\begin{aligned}
\Big|\Big((\varepsilon\boldsymbol{b}\cdot\nabla)(\boldsymbol{j}_h\boldsymbol{u} - \boldsymbol{u}), \boldsymbol{w}_h\Big)\Big| &= \big|\big(\boldsymbol{j}_h\boldsymbol{u} - \boldsymbol{u}, (\varepsilon\boldsymbol{b}\cdot\nabla)\boldsymbol{w}_h\big)\big| \\
&= \big|\big(\boldsymbol{j}_h\boldsymbol{u} - \boldsymbol{u}, \boldsymbol{\kappa}_h(\varepsilon\boldsymbol{b}\cdot\nabla)\boldsymbol{w}_h\big)\big| \\
&\leq C\sum_{K\in\mathcal{T}_h} h_K^{k+1}\|\boldsymbol{u}\|_{k+1,\omega(K)}\, \big\|\boldsymbol{\kappa}_h(\varepsilon\boldsymbol{b}\cdot\nabla)\boldsymbol{w}_h\big\|_{0,K}.
\end{aligned} \tag{3.44}$$

Now, let $\overline{\varepsilon\boldsymbol{b}}$ be the L^2-projection of $\varepsilon\boldsymbol{b}$ in the space of piecewise constant functions with respect to the decomposition $\mathcal{T}_h$. Using the L^2-stability of $\boldsymbol{\kappa}_h$, an inverse inequality, and $\boldsymbol{\kappa}_h(\overline{\varepsilon\boldsymbol{b}}\cdot\nabla)\boldsymbol{w}_h = \overline{\varepsilon\boldsymbol{b}}\cdot\boldsymbol{\kappa}_h(\nabla\boldsymbol{w}_h)$, we get for $\varepsilon\in W^{1,\infty}(\Omega)$

$$\begin{aligned}
\big\|\boldsymbol{\kappa}_h(\varepsilon\boldsymbol{b}\cdot\nabla)\boldsymbol{w}_h\big\|_{0,K} &\leq \big\|\boldsymbol{\kappa}_h((\varepsilon\boldsymbol{b} - \overline{\varepsilon\boldsymbol{b}})\cdot\nabla)\boldsymbol{w}_h\big\|_{0,K} + \big\|\boldsymbol{\kappa}_h(\overline{\varepsilon\boldsymbol{b}}\cdot\nabla)\boldsymbol{w}_h\big\|_{0,K} \\
&< C\, h_K|\varepsilon\boldsymbol{b}|_{1,\infty,K}\|\nabla\boldsymbol{w}_h\|_{0,K} + \|\boldsymbol{b}\|_{0,\infty,K}\, \big\|\boldsymbol{\kappa}_h(\nabla\boldsymbol{w}_h)\big\|_{0,K} \\
&\leq C\big\{|\varepsilon|_{1,\infty,K}\|\boldsymbol{b}\|_{0,\infty,K} + |\boldsymbol{b}|_{1,\infty,K}\big\}\|\boldsymbol{w}_h\|_{0,K} \\
&\quad + \|\boldsymbol{b}\|_{0,\infty,K}\, \big\|\boldsymbol{\kappa}_h(\nabla\boldsymbol{w}_h)\big\|_{0,K}.
\end{aligned} \tag{3.45}$$

Assuming $s > 0$, we conclude from (3.44)

$$\begin{aligned}
&\big|\big((\varepsilon\boldsymbol{b}\cdot\nabla)(\boldsymbol{j}_h\boldsymbol{u}-\boldsymbol{u}),\boldsymbol{w}_h\big)\big| \\
&\le C\sum_{K\in\mathcal{T}_h} h_K^{k+1}\|\boldsymbol{u}\|_{k+1,\omega(K)}\times \\
&\qquad\qquad \times\Big(\{|\varepsilon|_{1,\infty,K}\|\boldsymbol{b}\|_{0,\infty,K}+|\boldsymbol{b}|_{1,\infty,K}\}\,\|\boldsymbol{w}_h\|_{0,K}+\|\boldsymbol{b}\|_{0,\infty,K}\|\boldsymbol{\kappa}_h(\nabla\boldsymbol{w}_h)\|_{0,K}\Big) \\
&\le C\bigg(\sum_{K\in\mathcal{T}_h} h_K^{2k}\Big[h_K^2\|s\|_{0,\infty}^{-1}\,\{|\boldsymbol{b}|_{1,\infty,K}^2+|\varepsilon|_{1,\infty,K}^2\|\boldsymbol{b}\|_{0,\infty,K}^2\}+h_K^2\,\tau_K^{-1}\,\|\boldsymbol{b}\|_{0,\infty,K}^2\Big]\|\boldsymbol{u}\|_{k+1,\omega(K)}^2\bigg)^{1/2}\times \\
&\qquad \times\Big(\|\sqrt{s}\boldsymbol{w}_h\|_0^2+S_h\big((\boldsymbol{w}_h,0);(\boldsymbol{w}_h,0)\big)\Big)^{1/2}.
\end{aligned}$$

Analogously, we can estimate the next term by using (3.28)

$$\begin{aligned}
&\big|(p-j_hp,\nabla\cdot\boldsymbol{w}_h)\big|=\big|\big(p-j_hp,\kappa_h\nabla\cdot(\varepsilon\boldsymbol{w}_h)\big)\big| \\
&\le C\sum_{K\in\mathcal{T}_h} h_K^{k+1}\,\|p\|_{k+1,\omega(K)}\,\big\{|\varepsilon|_{1,\infty,K}\|\boldsymbol{w}_h\|_{0,K}+\|\kappa_h\nabla\cdot\boldsymbol{w}_h)\|_{0,K}\big\} \\
&\le C\bigg(\sum_{K\in\mathcal{T}_h} h_K^{2k}\,\Big[h_K^2\|s\|_{0,\infty}^{-1}|\varepsilon|_{1,\infty,K}^2+h_K^2\tau_K^{-1}\Big]\,\|p\|_{k+1,\omega(K)}^2\bigg)^{1/2}\times \\
&\qquad \times\Big(\|\sqrt{s}\boldsymbol{w}_h\|_0^2+S_h\big((\boldsymbol{w}_h,0);(\boldsymbol{w}_h,0)\big)\Big)^{1/2}.
\end{aligned}$$

Obviously, it holds

$$\begin{aligned}
\big|\big(r_h,\nabla\cdot\big(\varepsilon(\boldsymbol{j}_h\boldsymbol{u}-\boldsymbol{u})\big)\big)\big|&=\big|\big(\nabla r_h,\varepsilon\cdot(j_h\boldsymbol{u}-\boldsymbol{u})\big)\big|=\big|\big(\boldsymbol{\kappa}_h(\nabla r_h),\varepsilon\cdot(\boldsymbol{j}_h\boldsymbol{u}-\boldsymbol{u})\big)\big| \\
&\le C\bigg(\sum_{K\in\mathcal{T}_h} h_K^{2k+2}\,\alpha_K^{-1}\,\|\boldsymbol{u}\|_{k+1,\omega(K)}^2\bigg)^{1/2}\Big(S_h\big((0,r_h);(0,r_h)\big)\Big)^{1/2}.
\end{aligned}$$

Finally, we obtain

$$\begin{aligned}
&\big|S_h\big((\boldsymbol{j}_h\boldsymbol{u}-\boldsymbol{u},j_hp-p);(\boldsymbol{w}_h,r_h)\big)\big| \\
&\qquad\le\Big(S_h\big((\boldsymbol{j}_h\boldsymbol{u}-\boldsymbol{u},j_hp-p);(\boldsymbol{j}_h\boldsymbol{u}-\boldsymbol{u},j_hp-p)\big)\Big)^{1/2}\Big(S_h\big((\boldsymbol{w}_h,r_h);(\boldsymbol{w}_h,r_h)\big)\Big)^{1/2} \\
&\qquad\le C\bigg(\sum_{K\in\mathcal{T}_h} h_K^{2k}\Big[\big(\tau_K\,\|\boldsymbol{u}\|_{k+1,\omega(K)}^2+\alpha_K\,\|p\|_{k+1,\omega(K)}^2\Big]\bigg)^{1/2}|||(\boldsymbol{w}_h,r_h)|||.
\end{aligned}$$

Collecting all estimates above, we have shown

$$
\begin{aligned}
&|||(\boldsymbol{j}_h\boldsymbol{u}-\boldsymbol{u}_h, j_hp-p_h)||| \\
&\le C\Bigg(\sum_{K\in\mathcal{T}_h} h_K^{2k}\Big[Re^{-1}+h_K^2\,\|s\|_{0,\infty,K}+h_K^2\|s\|_{0,\infty,K}^{-1}\{|\boldsymbol{b}|_{1,\infty,K}^2+|\varepsilon|_{1,\infty,K}^2\|\boldsymbol{b}\|_{0,\infty,K}^2+1\} \\
&\qquad +h_K^2\,\tau_K^{-1}\|\boldsymbol{b}\|_{0,\infty,K}^2+h_K^2\,\alpha_K^{-1}+\tau_K+\alpha_K\Big]\Big(\|\boldsymbol{u}\|_{k+1,\omega(K)}^2+\|p\|_{k+1,\omega(K)}^2\Big)\Bigg)^{1/2}.
\end{aligned}
$$

By using the triangle inequality

$$
|||(\boldsymbol{u}-\boldsymbol{u}_h,p-p_h)|||\le|||(\boldsymbol{u}-\boldsymbol{j}_h\boldsymbol{u},p-j_hp)|||+|||(\boldsymbol{j}_h\boldsymbol{u}-\boldsymbol{u}_h,j_hp-p_h)|||
$$

and the approximation property

$$
\begin{aligned}
&|||(\boldsymbol{u}-\boldsymbol{j}_h\boldsymbol{u},p-j_hp)||| \\
&\le C\Bigg(\sum_{K\in\mathcal{T}_h} h_K^{2k}\Big[Re^{-1}+h_K^2\,\|s\|_{0,\infty,K}+(Re^{-1}+\|s\|_{0,\infty})\,h_K^2+\tau_K+\alpha_K\Big]\times \\
&\qquad\times\Big(\|\boldsymbol{u}\|_{k+1,\omega(K)}^2+\|p\|_{k+1,\omega(K)}^2\Big)\Bigg)^{1/2},
\end{aligned}
$$

we get

$$
\begin{aligned}
&|||(\boldsymbol{u}-\boldsymbol{u}_h,p-p_h)||| \\
&\le C\Bigg(\sum_{K\in\mathcal{T}_h} h_K^{2k}\Big[Re^{-1}+h_K^2\,\|s\|_{0,\infty,K}+h_K^2\|s\|_{0,\infty,K}^{-1}\{|\boldsymbol{b}|_{1,\infty,K}^2+|\varepsilon|_{1,\infty,K}^2\|\boldsymbol{b}\|_{0,\infty,K}^2+1\} \\
&\qquad +\tau_K^{-1}h_K^2\|\boldsymbol{b}\|_{0,\infty,K}^2+h_K^2\,\alpha_K^{-1}+\tau_K+\alpha_K\Big]\Big(\|\boldsymbol{u}\|_{k+1,\omega(K)}^2+\|p\|_{k+1,\omega(K)}^2\Big)\Bigg)^{1/2}
\end{aligned}
$$

which proves (3.39). Minimising the upper bound results in the choice $\tau_K \sim h_K$, and $\alpha_M \sim h_K$, which implies (3.40). □

3.5 Numerical results

3.5.1 Problem with smooth solution

We proceed our numerical investigations with solving a two dimensional problem (3.1) which is posed on the domain $\Omega = (0,1)^2$ and has the exact solution $(\boldsymbol{u}, p)$ from (2.125).

The porosity distribution is chosen as (2.126) and the convection field is prescribed by

$$\boldsymbol{b}(x,y) = \boldsymbol{u}(x,y) = \frac{1}{\varepsilon(x,y)} \begin{pmatrix} \sin(\pi x)\sin(\pi y) \\ \cos(\pi x)\cos(\pi y) \end{pmatrix}. \tag{3.46}$$

The right hand side $\boldsymbol{f}$ and boundary condition $\boldsymbol{g}$ are chosen such that (3.46) is the solution of the problem.

We apply equal order elements Q_k^+, $k = 1,2,3$ on cartesian meshes and choose for projection spaces P_0^{disc}, P_1^{disc} and P_2^{disc}, respectively. The coarse mesh consists of 2×2

Table 3.1: Total number of degrees of freedom (dof) for enriched spaces

level	dofs		
	Q_1^+	Q_2^+	Q_3^+
0	13	33	57
1	41	113	201
2	145	417	753
3	545	1,601	2,913
4	2,113	6,273	11,457
5	8,321	24,833	45,441
6	33,025	98,817	180,993

squares and will be uniformly refined. The corresponding numbers of degrees of freedom for one scalar solution component (velocity component or pressure) are shown in Table 3.1. We report errors for Oseen-like problem using a stronger triple-norm

$$|||(\boldsymbol{v},q)|||_* := \Big(Re^{-1}|\boldsymbol{v}|_1^2 + \|s\|_{0,\infty}\|\boldsymbol{v}\|_0^2 + (Re^{-1} + \|s\|_{0,\infty})\|q\|_0^2 + S_h\big((\boldsymbol{v},q);(\boldsymbol{v},q)\big)\Big)^{1/2}.$$

In our computations we set the Reynolds number to $Re = 1e+6$. The calculated rates of

Table 3.2: Oseen-like problem: LPS-error for $(Q_1^+, P_0^{\text{disc}})$ stabilisation

level	$\|\|\|(\boldsymbol{u}-\boldsymbol{u}_h, p-p_h)\|\|\|_*$	rate
0	4.078e+0	
1	1.304e+0	1.645
2	3.697e−1	1.818
3	1.271e−1	1.541
4	4.484e−2	1.503
5	1.585e−2	1.500
6	5.605e−3	1.500

convergence are in good agreement with theoretical results from Section 3.4, see Tables 3.2-3.4. The asymptotic behaviour of the LPS-error $|||(\boldsymbol{u}-\boldsymbol{u}_h, p-p_h)|||_*$ is shown in Figure 3.1.

Table 3.3: Oseen-like problem: LPS-error for $(Q_2^+, P_1^{\text{disc}})$ stabilisation

level	$\|\|\|(\boldsymbol{u}-\boldsymbol{u}_h, p-p_h)\|\|\|_*$	rate
0	1.284e+0	
1	3.051e−1	2.074
2	5.514e−2	2.468
3	9.529e−3	2.533
4	1.589e−3	2.584
5	2.602e−4	2.611
6	4.127e−5	2.656

Table 3.4: Oseen-like problem: LPS-error for $(Q_3^+, P_2^{\text{disc}})$ stabilisation

level	$\|\|\|(\boldsymbol{u}-\boldsymbol{u}_h, p-p_h)\|\|\|_*$	rate
0	7.147e−1	
1	6.933e−2	3.366
2	6.162e−3	3.492
3	5.452e−4	3.498
4	4.800e−5	3.506
5	4.233e−6	3.503
6	3.738e−7	3.501

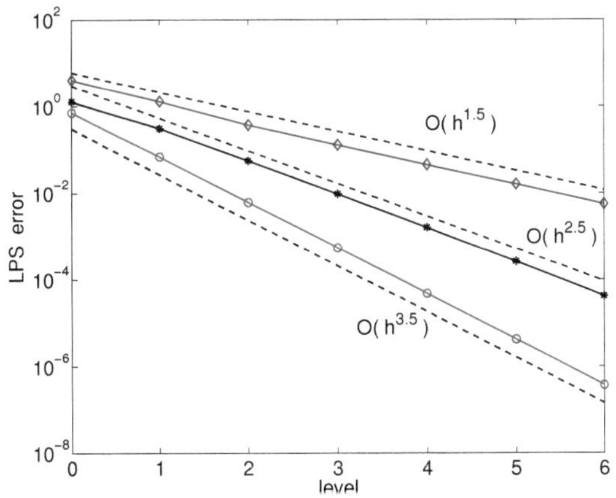

Figure 3.1: Oseen-like problem: LPS error.

4 Enhancing accuracy of numerical solution

Several types of superconvergence in finite element methods, as indicated by Brandts and Křížek [13], have been studied in the last two decades. We consider a superconvergence property of post-processing type [57] which increases the order of convergence of the original finite element solution in case of the Stokes-like and Brinkman–Forchheimer problem, respectively. Such a post-processing is nothing but a higher order interpolation on a coarser mesh of the original finite element solution. For proving the superconvergence property we need two main ingredients:

- An interpolation (in the same finite element space) approximating the finite element solution of higher order. Often such an interpolation does exist if the underlying mesh has a special construction, cf. [67]. This special interpolation is called to satisfy a superclose property, cf. [80].
- A higher order interpolation of the original finite element solution to achieve higher order accuracy. The interpolated finite element solution resulting from the post-processing is called to satisfy a superconvergence property, cf. [13].

The two steps above have been examined for many different conforming finite elements including mixed finite elements (cf. [53, 56]). For nonconforming finite elements applied to the Poisson equation we refer to [58].
The superclose phenomena for the Stokes problem in the two-dimensional case have been already reported in [54, 55]. In [54] the $(4 \times Q_1, Q_1)$ element has been studied whereas in [55] the Bernardi–Raugel element [7] and the (Q_2, P_1^{disc}) element are considered. The estimates in both papers are based on exact integral identities where some details are missing. Unfortunately, the post-processing operators in [55] are not unisolvent and cannot be used to derive a superconvergence result. This lack has been overcome in [61]. The superconvergence results have been stated therein for Stokes and stationary Navier–Stokes problems in three dimensions by using (Q_2, P_1^{disc}) conforming elements.

4.1 Superconvergence of finite elements applied to Brinkman–Forchheimer problem

In this section we recall the general principle of superconvergence applied to saddle point problems for conforming finite element pairs satisfying the discrete inf-sup condition.

We denote by $\mathcal{T}_h$ a shape regular decomposition of $\Omega = (0,1)^2$ into (open) cells K such that

$$\overline{\Omega} = \bigcup_{K \in \mathcal{T}_h} \overline{K}.$$

We denote by $\boldsymbol{i}_h : \boldsymbol{H}^2(\Omega) \to \boldsymbol{X}_h$ and $j_h : M \to M_h$ appropriate interpolation operators. We assume that $\boldsymbol{i}_h \boldsymbol{u}|_{\partial\Omega}$ depends on $\boldsymbol{g} = \boldsymbol{u}|_{\partial\Omega}$ only and set $\boldsymbol{g}_h := \boldsymbol{i}_h \boldsymbol{u}|_{\partial\Omega}$.

We start our analysis with the following discrete stability result from [30]: There is a positive constant C independent of h such that

$$\frac{\varepsilon_0}{Re} |\boldsymbol{w}_h|_1 + \|r_h\|_0 \leq C \sup_{\substack{(\boldsymbol{v}_h, q_h) \in \boldsymbol{X}_{h0} \times M_h \\ (\boldsymbol{v}_h, q_h) \neq (\boldsymbol{0}, 0)}} \frac{a(\boldsymbol{w}_h, \boldsymbol{v}_h) + c(\boldsymbol{w}_h, \boldsymbol{v}_h) - b(\boldsymbol{v}_h, r_h) + b(\boldsymbol{w}_h, q_h)}{|\boldsymbol{v}_h|_1 + \|q_h\|_0},$$

holds for all $(\boldsymbol{w}_h, r_h) \in \boldsymbol{X}_{h0} \times M_h$ see also [72, Chapter 5.1] for detailed proof for abstract saddle point problems. Setting $\boldsymbol{w}_h := \boldsymbol{u}_h - \boldsymbol{i}_h \boldsymbol{u} \in \boldsymbol{X}_{h0}$, $r_h := p_h - j_h p \in M_h$ and using the Galerkin orthogonality

$$\begin{aligned} 0 = &\, a(\boldsymbol{u}_h - \boldsymbol{u}, \boldsymbol{v}_h) + c(\boldsymbol{u}_h - \boldsymbol{u}, \boldsymbol{v}_h) - b(\boldsymbol{v}_h, p_h - p) + b(\boldsymbol{u}_h - \boldsymbol{u}, q_h) \\ &+ \tilde{n}(\boldsymbol{u}_h, \boldsymbol{u}_h, \boldsymbol{v}_h) - \tilde{n}(\boldsymbol{u}, \boldsymbol{u}, \boldsymbol{v}) \\ &+ d(\boldsymbol{u}_h; \boldsymbol{u}_h, \boldsymbol{v}_h) - d(\boldsymbol{u}; \boldsymbol{u}, \boldsymbol{v}) \qquad \forall (\boldsymbol{v}_h, q_h) \in \boldsymbol{X}_{h0} \times M_h, \end{aligned} \tag{4.1}$$

we obtain

$$\begin{aligned} &\frac{\varepsilon_0}{Re} |\boldsymbol{u}_h - \boldsymbol{i}_h \boldsymbol{u}|_1 + \|p_h - j_h p\|_0 \\ &\leq C \sup_{\substack{(\boldsymbol{v}_h, q_h) \in \boldsymbol{X}_{h0} \times M_h \\ (\boldsymbol{v}_h, q_h) \neq (\boldsymbol{0}, 0)}} \frac{E(\boldsymbol{u}, p; \boldsymbol{v}_h, q_h) + \tilde{n}(\boldsymbol{u}, \boldsymbol{u}, \boldsymbol{v}_h) - \tilde{n}(\boldsymbol{u}_h, \boldsymbol{u}_h, \boldsymbol{v}_h) + d(\boldsymbol{u}; \boldsymbol{u}, \boldsymbol{v}_h) - d(\boldsymbol{u}_h; \boldsymbol{u}_h, \boldsymbol{v}_h)}{|\boldsymbol{v}_h|_1 + \|q_h\|_0} \end{aligned} \tag{4.2}$$

with

$$\begin{aligned} E(\boldsymbol{u}, p; \boldsymbol{v}_h, q_h) := &\, a(\boldsymbol{u} - \boldsymbol{i}_h \boldsymbol{u}, \boldsymbol{v}_h) + c(\boldsymbol{u} - \boldsymbol{i}_h \boldsymbol{u}, \boldsymbol{v}_h) \\ &- b(\boldsymbol{v}_h, p - j_h p) + b(\boldsymbol{u} - \boldsymbol{i}_h \boldsymbol{u}, q_h). \end{aligned} \tag{4.3}$$

Remark 4.1 *Standard error estimates are based on the continuity of* $E(\boldsymbol{u}, p; \cdot, \cdot)$ *on* $\boldsymbol{X} \times M$ *resulting in*

$$|E(\boldsymbol{u}, p; \boldsymbol{v}_h, q_h)| \leq C \left(|\boldsymbol{u} - \boldsymbol{i}_h \boldsymbol{u}|_1 + \|p - j_h p\|_0 \right) \left(|\boldsymbol{v}_h|_1 + \|q_h\|_0 \right). \tag{4.4}$$

For a pair of finite element spaces of k-th order, which means that there are interpolation operators $\boldsymbol{i}_h$ and j_h such that

$$|\boldsymbol{u} - \boldsymbol{i}_h \boldsymbol{u}|_1 + \|p - j_h p\|_0 \leq C\, h^k \left(|\boldsymbol{u}|_{k+1} + |p|_k\right),$$

we conclude

$$|E(\boldsymbol{u}, p; \boldsymbol{v}_h, q_h)| \leq C\, h^k \left(|\boldsymbol{u}|_{k+1} + |p|_k\right)\left(|\boldsymbol{v}_h|_1 + \|q_h\|_0\right).$$

However, in the next section we will show that for a special finite element pair of order k and suitable chosen interpolation operators the estimate

$$|E(\boldsymbol{u}, p; \boldsymbol{v}_h, q_h)| \leq C\, h^{k+1} \left(|\boldsymbol{u}|_{k+2} + |p|_{k+1}\right)\left(|\boldsymbol{v}_h|_1 + \|q_h\|_0\right). \tag{4.5}$$

can be established for all $(\boldsymbol{v}_h, q_h) \in \boldsymbol{X}_{h0} \times M_h$. Thus, in view of (4.2) the supercloseness of the discrete solution $(\boldsymbol{u}_h, p_h)$ to the interpolated solution $(\boldsymbol{i}_h\boldsymbol{u}, j_h p)$ can be shown if it holds also

$$\left|\tilde{n}(\boldsymbol{u}, \boldsymbol{u}, \boldsymbol{v}_h) - \tilde{n}(\boldsymbol{u}_h, \boldsymbol{u}_h, \boldsymbol{v}_h) + d(\boldsymbol{u}; \boldsymbol{u}, \boldsymbol{v}_h) - d(\boldsymbol{u}_h; \boldsymbol{u}_h, \boldsymbol{v}_h)\right| \leq C h^{k+1} \tag{4.6}$$

Now we consider a coarser decomposition $\mathcal{T}_{2h}$ of Ω into patches P such that

$$\overline{\Omega} = \bigcup_{P \in \mathcal{T}_{2h}} \overline{P}$$

and each closed patch $\overline{P}$ consists of a fixed number of closed cells $\overline{K}$. Often the decomposition $\mathcal{T}_h$ can be generated from $\mathcal{T}_{2h}$ by a regular refinement, i.e., a patch consists of 8 cells. On this new decomposition, we introduce finite element spaces $\boldsymbol{Y}_{2h}$ and N_{2h} and corresponding interpolation operators $\boldsymbol{I}_{2h} : \boldsymbol{C}(\overline{\Omega}) \to \boldsymbol{Y}_{2h}$ and $J_{2h} : M \to N_{2h}$, respectively. For a pair of k-th order finite element spaces $\boldsymbol{X}_h$, M_h we assume that the following conditions are satisfied:

$$\begin{array}{lll}
\text{(A)} & \boldsymbol{I}_{2h}\boldsymbol{i}_h\boldsymbol{u} = \boldsymbol{I}_{2h}\boldsymbol{u} & \forall \boldsymbol{u} \in \boldsymbol{C}(\overline{\Omega}), \\
 & J_{2h} j_h p = J_{2h} p & \forall p \in M, \\
\text{(B)} & |\boldsymbol{u} - \boldsymbol{I}_{2h}\boldsymbol{u}|_1 \leq C\, h^{k+1} |\boldsymbol{u}|_{k+2} & \forall \boldsymbol{u} \in \boldsymbol{H}^{k+2}(\Omega), \\
 & \|p - J_{2h} p\|_0 \leq C\, h^{k+1} |p|_{k+1} & \forall p \in H^{k+1}(\Omega), \\
\text{(C)} & |\boldsymbol{I}_{2h}\boldsymbol{u}_h|_1 \leq C\, |\boldsymbol{u}_h|_1 & \forall \boldsymbol{u}_h \in \boldsymbol{X}_h, \\
 & \|J_{2h} p_h\|_0 \leq C\, \|p_h\|_0 & \forall p_h \in M_h.
\end{array}$$

Then, we can also derive a superconvergence result for the error of the post-processed solution $(\boldsymbol{I}_{2h}\boldsymbol{u}_h, J_{2h}p_h)$ to the solution $(\boldsymbol{u}, p)$ of (2.8) which is assumed to belong to $\boldsymbol{H}^{k+2}(\Omega) \times H^{k+1}(\Omega)$. More precisely, the estimates

$$\begin{aligned}
|\boldsymbol{u} - \boldsymbol{I}_{2h}\boldsymbol{u}_h|_1 &\leq |\boldsymbol{u} - \boldsymbol{I}_{2h}\boldsymbol{i}_h\boldsymbol{u}|_1 + |\boldsymbol{I}_{2h}\boldsymbol{i}_h\boldsymbol{u} - \boldsymbol{I}_{2h}\boldsymbol{u}_h|_1 \\
&\leq |\boldsymbol{u} - \boldsymbol{I}_{2h}\boldsymbol{u}|_1 + C\, |\boldsymbol{i}_h\boldsymbol{u} - \boldsymbol{u}_h|_1, \\
|\boldsymbol{u} - \boldsymbol{I}_{2h}\boldsymbol{u}_h|_1 &\leq C\, h^{k+1} \left(|\boldsymbol{u}|_{k+2} + |p|_{k+1}\right)
\end{aligned} \tag{4.7}$$

and

$$
\begin{aligned}
\|p - J_{2h}p_h\|_0 &\le \|p - J_{2h}j_hp\|_0 + \|J_{2h}j_hp - J_{2h}p_h\|_0 \\
&\le \|p - J_{2h}p\|_0 + C\,\|j_hp - p_h\|_0, \\
\|p - J_{2h}p_h\|_0 &\le C\,h^{k+1}\,(|\boldsymbol{u}|_{k+2} + |p|_{k+1}) \qquad (4.8)
\end{aligned}
$$

hold true. Note that in general all interpolation operators are nonstandard.

4.2 Supercloseness of the (Q_2, P_1^{disc}) element

In this section we specify the finite element spaces $\boldsymbol{X}_h$ and M_h and prove an estimate of type (4.5) for $k = 2$ and appropriate interpolation operators on a family of quadrilateral meshes. We assume that the edges of each cell K are parallel to the coordinate axes and that K is a rectangle with sides of length $2h_{x,K}$, $2h_{y,K}$. We suppose that the family of meshes is shape regular in the following sense: there is a positive constant C such that

$$C\sqrt{h_{x,K}^2 + h_{y,K}^2} \le \min(h_{x,K}, h_{y,K}) \qquad \forall K \in \mathcal{T}_h.$$

Furthermore, we set $h_K = \operatorname{diam} K = 2\sqrt{h_{x,K}^2 + h_{y,K}^2}$ and $h = \max_{K\in\mathcal{T}_h}\{h_K\}$.

We use the space of continuous, piecewise biquadratic functions for the velocity, and and the space of discontinuous, piecewise linear functions having mean value zero for the pressure. From Theorem 2.20 it is known that this finite element pair fulfils the discrete inf-sup condition (2.48). Due to the interpolation properties of the standard interpolation operators we get from (4.4)

$$|E(\boldsymbol{u}, p; \boldsymbol{v}_h, q_h)| \le C\,h^2\,(|\boldsymbol{u}|_3 + |p|_2)\,(|\boldsymbol{v}_h|_1 + \|q_h\|_0)\,.$$

We want to show that for nonstandard interpolation operators a stronger result can be obtained.

The interpolation operators $\boldsymbol{i}_h : \boldsymbol{H}^2(\Omega) \to \boldsymbol{X}_h$ and $j_h : M \to M_h$ are locally defined on each cell K by mapping K onto the reference cell $\widehat{K} = (-1,+1)^2$ via the affine, bijective mapping $\boldsymbol{F}_K^{-1} : K \to \widehat{K}$. We denote the 4 vertices of $\widehat{K}$ by $\hat{\boldsymbol{a}}_i$, $i = 1,\dots,4$, the 4 edges of $\widehat{K}$ by $\hat{l}_i$, $i = 1,\dots,4$. On the reference cell $\widehat{K}$ we define first a scalar interpolation operator $i_{\widehat{K}} : H^2(\widehat{K}) \to Q_2$ by the 9 nodal functionals

$$n_i(\hat{v}) = \hat{v}(\boldsymbol{a}_i),\quad i = 1,\dots,4,\quad n_{i+4}(\hat{v}) = \frac{1}{2}\int_{\hat{l}_i} \hat{v}\,d\hat{s},\quad i = 1,\dots,4,\quad n_9(\hat{v}) = \frac{1}{4}\int_{\widehat{K}} \hat{v}\,d\xi\,d\eta,$$

such that

$$n_j(i_{\hat K}\hat v) = n_j(\hat v), \qquad j = 1,\dots,9.$$

Now the scalar interpolation $i_h : H^2(\Omega) \to X_h$ is piecewise defined by

$$i_h(v)\Big|_K := (i_{\widehat K}(v|_K \circ \boldsymbol{F}_K)) \circ \boldsymbol{F}_K^{-1},$$

and for $\boldsymbol{v} = (v_1, v_2) \in \boldsymbol{H}^2(\Omega)$ the vector-valued interpolation $\boldsymbol{i}_h\boldsymbol{v}$ is given by

$$\boldsymbol{i}_h(\boldsymbol{v}) := (i_h(v_1), i_h(v_2)).$$

Note that the piecewise defined interpolation $i_h(v)$ fits to a global continuous function because the restriction of a function of Q_2 onto a face of $\widehat K$ is a quadratic function of one variable which is uniquely defined by the subset of 2 vertex and 1 edge nodal functionals living on this edge. In addition the property that $\boldsymbol{i}_h\boldsymbol{u}|_{\partial\Omega}$ depends only on $\boldsymbol{u}|_{\partial\Omega}$ is satisfied.

The nodal functionals for the pressure space on the reference element $\hat K$ are defined by

$$m_i(\hat q) := \frac{1}{4}\int_{\widehat K} \hat q\, r_i \, d\xi d\eta \qquad i = 1,\dots,3, \tag{4.9}$$

where $r_1 = 1$, $r_2 = \xi$, $r_3 = \eta$. The canonical interpolation $j_{\widehat K} : L^2(\widehat K) \to P_1$ is the L^2-projection given by

$$m_i(j_{\widehat K}\hat q) = m_i(\hat q), \qquad i = 1,\dots,3\,,$$

resulting into the global interpolation $j_h : M \to M_h$ with

$$j_h(q)\Big|_K = \big((j_{\widehat K}(q \circ \boldsymbol{F}_K)\big) \circ \boldsymbol{F}_K^{-1}.$$

In the following we shall use the abbreviations

$$\partial_x W := \{\partial_x w \,:\, w \in W\} \quad \text{and} \quad \partial_y W := \{\partial_y w \,:\, w \in W\}\,.$$

Lemma 4.1 *Let $\varepsilon \in W^{1,\infty}(\Omega)$, $\boldsymbol{u} \in \boldsymbol{H}^4(\Omega)$ and let $\boldsymbol{i}_h\boldsymbol{u}$ be the interpolant defined above. Then, on a family of rectangular meshes we have*

$$\big|\big(\varepsilon\nabla(\boldsymbol{u} - \boldsymbol{i}_h\boldsymbol{u}), \nabla\boldsymbol{v}_h\big)\big| \le Ch^3(|\boldsymbol{u}|_3 + |\boldsymbol{u}|_4)\,|\boldsymbol{v}_h|_1 \quad \forall \boldsymbol{v}_h \in \boldsymbol{X}_h \tag{4.10}$$

$$\big|\big(div\,(\varepsilon(\boldsymbol{u} - \boldsymbol{i}_h\boldsymbol{u})), q_h\big)\big| \le Ch^3(|\boldsymbol{u}|_3 + |\boldsymbol{u}|_4)\,\|q_h\|_0 \quad \forall q_h \in M_h. \tag{4.11}$$

$$\big|\big(\alpha(\boldsymbol{u} - \boldsymbol{i}_h\boldsymbol{u}), \boldsymbol{v}_h\big)\big| \le Ch^3|\boldsymbol{u}|_3\,|\boldsymbol{v}_h|_1 \quad \forall \boldsymbol{v}_h \in \boldsymbol{X}_{h0} \tag{4.12}$$

Proof. Let $\bar\varepsilon$ be the L^2-projection of ε in the space of piecewise constant functions with respect to the decomposition $\mathcal{T}_h$. We set

$$\bar\varepsilon|_K = \varepsilon_K \in \mathbb{R} \qquad \forall\, K \in \mathcal{T}_h\,. \tag{4.13}$$

Furthermore, let us assume $\varepsilon \in W^{1,\infty}(\Omega)$. We have

$$\begin{aligned}&\left|\left(\varepsilon\nabla(\boldsymbol{u} - \boldsymbol{i}_h\boldsymbol{u}), \nabla\boldsymbol{v}_h\right)\right| \\ &\leq \sum_{K\in\mathcal{T}_h}\left|\left((\varepsilon - \varepsilon_K)\nabla(\boldsymbol{u} - \boldsymbol{i}_h\boldsymbol{u}), \nabla\boldsymbol{v}_h\right)_K\right| + \sum_{K\in\mathcal{T}_h}|\varepsilon_K|\left|\left(\nabla(\boldsymbol{u} - \boldsymbol{i}_h\boldsymbol{u}), \nabla\boldsymbol{v}_h\right)_K\right|.\end{aligned} \tag{4.14}$$

The first sum from the above splitting can be bounded by using Hölder inequality, Bramble–Hilbert lemma and interpolation estimate

$$\begin{aligned}\left|\left((\varepsilon - \varepsilon_K)\nabla(\boldsymbol{u} - \boldsymbol{i}_h\boldsymbol{u}), \nabla\boldsymbol{v}_h\right)_K\right| &\leq |\varepsilon - \varepsilon_K|_{0,\infty,K}|\boldsymbol{u} - \boldsymbol{i}_h\boldsymbol{u}|_{1,K}\,|\boldsymbol{v}_h|_{1,K} \\ &\leq C h_K^3\,|\varepsilon|_{1,\infty,K}|\boldsymbol{u}|_{3,K}\,|\boldsymbol{v}_h|_{1,K}\end{aligned} \tag{4.15}$$

The second sum from (4.14) can be estimated using the fact $|\varepsilon_K| \leq \|\varepsilon\|_{0,\infty,K}$ and employing the supercloseness of $\boldsymbol{i}_h$, see [74, 72, 57] ($n = 2$) and [61] ($n = 3$). To this end, we consider the scalar case and for each mesh rectangle K define $E_K, F_K : K \to \mathbb{R}$ with

$$\begin{aligned}E_K(x) &= \frac{1}{2}\left[(x - x_K)^2 - h_{x,K}^2\right], \\ F_K(y) &= \frac{1}{2}\left[(y - y_K)^2 - h_{y,K}^2\right],\end{aligned} \tag{4.16}$$

whereby (x_K, y_K) denotes the barycentre of K, see Figure 4.1.

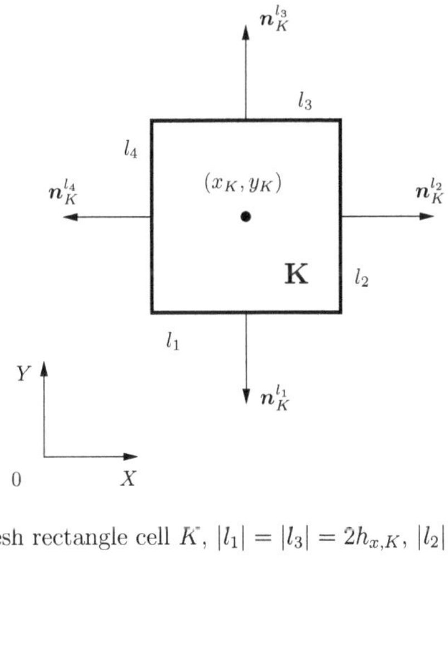

Figure 4.1: Mesh rectangle cell K, $|l_1| = |l_3| = 2h_{x,K}$, $|l_2| = |l_4| = 2h_{y,K}$

After [57] it holds

$$\begin{aligned}\big(\partial_x(w-i_hw),\partial_x v_h\big)_K = &-\frac{1}{30}\int\limits_K F_K^2(y)F_K'(y)\partial_{xyyy}w(x,y)\partial_{xyy}v_h(x,y_K)\,dxdy\\ &-\frac{1}{6}\int\limits_K F_K^2(y)\partial_{xyyy}w(x,y)\partial_{xy}v_h(x,y_K)\,dxdy\end{aligned} \tag{4.17}$$

for all $w\in H^4(K)$ and $v_h\in Q_2(K)$. Using the facts that

$$|F_K^2(y)| \le \left(\frac{h_{y,K}^2}{2}\right)^2 \quad \text{and} \quad |F_K'(y)| \le h_{y,K}\,, \tag{4.18}$$

$$\partial_{xyy}v_h(x,y_K) = \partial_{xyy}v_h(x,y) \qquad \forall\, v_h\in Q_2(K)\,,$$

$$\partial_{xy}v_h(x,y_K) = \partial_{xy}v_h(x,y) - (y-y_K)\partial_{xyy}v_h(x,y) \qquad \forall\, v_h\in Q_2(K)$$

as well as the inverse inequalities, we get the superclose bound

$$|\big(\partial_x(w-i_hw),\partial_x v_h\big)_K| \le Ch_{y,K}^3\|\partial_{xyyy}w\|_{0,K}\,\|\partial_x v_h\|_{0,K}\,, \tag{4.19}$$

and by analogy

$$|\big(\partial_y(w-i_hw),\partial_y v_h\big)_K| \le Ch_{x,K}^3\|\partial_{xxxy}w\|_{0,K}\,\|\partial_y v_h\|_{0,K}\,. \tag{4.20}$$

Therefore, we obtain

$$\big(\nabla(\boldsymbol{u}-\boldsymbol{i}_h\boldsymbol{u}),\nabla\boldsymbol{v}_h\big)_K \le Ch_K^3|\boldsymbol{u}|_{4,K}\,|\boldsymbol{v}_h|_{1,K}\,. \tag{4.21}$$

Collecting (4.14)-(4.21), we deduce (4.10). Now, we have

$$\big(\mathrm{div}\,(\varepsilon(\boldsymbol{u}-\boldsymbol{i}_h\boldsymbol{u})),q_h\big) = \big(\boldsymbol{u}-\boldsymbol{i}_h\boldsymbol{u},q_h\nabla\varepsilon\big) + \big(\varepsilon\,\mathrm{div}\,(\boldsymbol{u}-\boldsymbol{i}_h\boldsymbol{u}),q_h\big)\,. \tag{4.22}$$

Applying Hölder inequality and interpolation estimate, we get

$$\big|\big(\boldsymbol{u}-\boldsymbol{i}_h\boldsymbol{u},q_h\nabla\varepsilon\big)_K\big| \le C|\varepsilon|_{1,\infty,K}h_K^3|\boldsymbol{u}|_{3,K}\|q_h\|_{0,K}\,,$$

and consequently

$$\big|\big(\boldsymbol{u}-\boldsymbol{i}_h\boldsymbol{u},q_h\nabla\varepsilon\big)\big| \le Ch^3|\boldsymbol{u}|_3\|q_h\|_0\,. \tag{4.23}$$

Now, we get from the triangle inequality

$$\big|\big(\varepsilon\,\mathrm{div}\,(\boldsymbol{u}-\boldsymbol{i}_h\boldsymbol{u}),q_h\big)\big| \le \big|\big(\varepsilon\,\partial_x\,(u_1-i_hu_1),q_h\big)\big| + \big|\big(\varepsilon\,\partial_y\,(u_2-i_hu_2),q_h\big)\big|\,.$$

Approximating ε by piecewise constants with respect to the triangulation $\mathcal{T}_h$, taking into consideration the fact

$$q_h|_K \in P_1 \subset \partial_x Q_2 \cap \partial_y Q_2$$

and using the local estimates (4.19) and (4.20), we conclude by analogy to the proof of (4.10) the global bound

$$\left|\left(\varepsilon\,\mathrm{div}\,(\boldsymbol{u}-\boldsymbol{i}_h\boldsymbol{u}), q_h\right)\right| \le Ch^3(|\boldsymbol{u}|_3+|\boldsymbol{u}|_4)\,\|q_h\|_0\,. \tag{4.24}$$

Collecting (4.22)-(4.24) gives the assertion (4.11). Finally, the Hölder inequality, interpolation estimate imply the local bound

$$\begin{aligned}\left|\left(\alpha(\boldsymbol{u}-\boldsymbol{i}_h\boldsymbol{u}), \boldsymbol{v}_h\right)_K\right| &\le C\|\alpha\|_{0,\infty,K}\|\boldsymbol{u}-\boldsymbol{i}_h\boldsymbol{u}\|_{0,K}\|\boldsymbol{v}_h\|_{0,K}\\ &\le C\|\alpha\|_{0,\infty,K}h_K^3|\boldsymbol{u}|_{3,K}\|\boldsymbol{v}_h\|_{0,K}\,.\end{aligned} \tag{4.25}$$

Employing Poincaré inequality, yields the global bound (4.12). □

Next, we bound the term $b(\boldsymbol{v}_h, p-j_hp)$. To this end, we assume that the mesh is quasi uniform in each coordinate direction, i.e., for two cells K and K' with a common face let

$$\max\left(|h_K-h_{K'}|, |k_K-k_{K'}|\right) \le C\,h^2\,. \tag{4.26}$$

Lemma 4.2 *Let $p \in H^3(\Omega)$, $\varepsilon \in W^{2,\infty}(\Omega)$ and let the mesh be quasi uniform in each coordinate direction. Then, we have for each $\boldsymbol{v}_h \in \boldsymbol{X}_{h0}$*

$$|(\mathit{div}\,(\varepsilon\boldsymbol{v}_h), p-j_hp)| \le Ch^3\,\|p\|_3|\boldsymbol{v}_h|_1. \tag{4.27}$$

Proof. First, we observe

$$(\mathrm{div}\,(\varepsilon\boldsymbol{v}_h), p-j_hp) = (\nabla\varepsilon\cdot\boldsymbol{v}_h, p-j_hp) + (\varepsilon\,\mathrm{div}\,(\boldsymbol{v}_h), p-j_hp)\,. \tag{4.28}$$

Using the fact that due to (4.9) j_h is L_2 projection onto M_h, employing Cauchy–Schwarz inequality, interpolation estimates, we obtain

$$\begin{aligned}|(\nabla\varepsilon\cdot\boldsymbol{v}_h, p-j_hp)_K| &= \left|\left(\nabla\varepsilon\cdot\boldsymbol{v}_h - j_h(\nabla\varepsilon\cdot\boldsymbol{v}_h), p-j_hp\right)_K\right|\\ &\le \|\nabla\varepsilon\cdot\boldsymbol{v}_h - j_h(\nabla\varepsilon\cdot\boldsymbol{v}_h)\|_{0,K}\,\|p-j_hp\|_{0,K}\\ &\le Ch_K^3|\nabla\varepsilon\cdot\boldsymbol{v}_h|_{1,K}|p|_{2,K}\\ &\le Ch_K^3\|\varepsilon\|_{2,\infty,K}|p|_{2,K}\|\boldsymbol{v}_h\|_{1,K}\,.\end{aligned}$$

From Poincaré inequality follows then the global bound

$$|(\nabla\varepsilon\cdot\boldsymbol{v}_h, p-j_hp)| \le Ch^3\|\varepsilon\|_{2,\infty}|p|_2|\boldsymbol{v}_h|_1\,. \tag{4.29}$$

Next, we get for ε_K from (4.13)

$$\begin{aligned}&(\varepsilon\,\mathrm{div}\,(\boldsymbol{v}_h), p-j_hp)\\ &= \sum_{K\in\mathcal{T}_h}\left((\varepsilon-\varepsilon_K)\mathrm{div}\,(\boldsymbol{v}_h), p-j_hp\right)_K + \sum_{K\in\mathcal{T}_h}\varepsilon_K\left(\mathrm{div}\,(\boldsymbol{v}_h), p-j_hp\right)_K\,.\end{aligned} \tag{4.30}$$

The first sum can be estimated by employing locally Hölder-inequality, Bramble-Hilbert lemma and interpolation estimate

$$\left|\left((\varepsilon-\varepsilon_K)\mathrm{div}\,(\boldsymbol{v}_h),p-j_hp\right)_K\right| \leq Ch_K^3|\varepsilon|_{1,\infty,K}|p|_{2,K}|\boldsymbol{v}_h|_{1,K}\,. \tag{4.31}$$

Let us recall the supercloseness result from [72] ($n=2$) in order to get a bound for the second sum from (4.30)

$$\begin{aligned}
(\bar\varepsilon\partial_x v_{h1},p-j_hp) &= \sum_{K\in\mathcal{T}_h}\varepsilon_K\Big[\\
&-\frac{1}{36}\int\limits_K\{\left((F_K^2)'(E_K^2)''-2h_{x,K}^2(F_K^2)'\right)\partial_{xyy}p-2h_{y,K}^2(E_K^2)'\partial_{xxy}p\}\partial_{xxy}v_{h1}\,dxdy \\
&+\frac{1}{36}\int\limits_K\{\left((F_K^2)'(E_K^2)''-2h_{x,K}^2(F_K^2)'\right)\partial_{xyy}p-2h_{y,K}^2(E_K^2)'\partial_{xxy}p\}F_K'\partial_{xxyy}v_{h1}\,dxdy \\
&-\frac{4}{36}h_{x,K}^2h_{y,K}^2\int\limits_K\partial_{xyy}p\partial_{xx}v_{h1}dxdy \\
&+\frac{4}{36}h_{x,K}^2h_{y,K}^2\int\limits_{l_3}\partial_{xy}p\partial_{xx}v_{h1}dx-\frac{4}{36}h_{x,K}^2h_{y,K}^2\int\limits_{l_1}\partial_{xy}p\partial_{xx}v_{h1}dx \\
&+\frac{4}{36}h_{x,K}^2h_{y,K}^2\int\limits_K F_K\partial_{xyy}p\partial_{xxyy}v_{h1}dxdy \\
&-\frac{1}{6}\int\limits_K\partial_{xyy}pF_K^2\partial_{yy}v_{h1}dxdy \\
&+\frac{1}{6}\int\limits_{l_2}\partial_{yy}pF_K^2\partial_{yy}v_{h1}dy-\frac{1}{6}\int\limits_{l_4}\partial_{yy}pF_K^2\partial_{yy}v_{h1}dy\Big]
\end{aligned} \tag{4.32}$$

whereby v_{h1} stands the first component of $\boldsymbol{v}_h\in\boldsymbol{X}_{h0}$ and l_1, l_3 and l_2, l_4 denote horizontal and vertical edges of the cell K, respectively (see Figure 4.1). The formula (4.32) can be derived using the expansion of $v_{h1}\in Q_2(K)$

$$\begin{aligned}
\partial_x v_{h1}(x,y) &= \partial_x v_{h1}(x,y_K)+F_K'\partial_{xy}v_{h1}(x_K,y_K)+E_K'F_K'\partial_{xxy}v_{h1}(x,y_K) \\
&+\frac{1}{6}\left((F_K^2)''+h_{y,K}^2\right)\partial_{xyy}v_{h1}(x,y)\,,
\end{aligned}$$

integrating by parts and employing the properties of the pressure interpolation operator j_h and functions E_K, F_K. The estimates of the cell integrals from the above sum follow from the Hölder and inverse inequalities and properties of functions E_K, F_K and ε, e.g.,

$$\begin{aligned}
&\left| -\frac{\varepsilon_K}{36} \int\limits_K \{ ((F_K^2)'(E_K^2)'' - 2h_{x,K}^2 (F_K^2)')\partial_{xyy}p - 2h_{y,K}^2 (E_K^2)'\partial_{xxy}p \} \partial_{xxy} v_{h1}\, dxdy \right| \\
&\le C\|\varepsilon\|_{0,\infty,K} \left[\|(F_K^2)'\|_{0,\infty,K} \|(E_K^2)''\|_{0,\infty,K} + 2h_{x,K}^2 \|(F_K^2)'\|_{0,\infty,K} \right] \|\partial_{xyy}p\|_{0,K} \|\partial_{xxy}v_{h1}\|_{0,K} \\
&\quad + C\|\varepsilon\|_{0,\infty,K}\, h_{y,K}^2 \|(E_K^2)'\|_{0,\infty,K} \|\partial_{xxy}p\|_{0,K} \|\partial_{xxy}v_{h1}\|_{0,K} \\
&\le C\|\varepsilon\|_{0,\infty,K} (h_{y,K}^3 h_{x,K}^2 + h_{x,K}^2 h_{y,K}^3) \|\partial_{xyy}p\|_{0,K} h_{x,K}^{-1} h_{y,K}^{-1} \|\partial_x v_{h1}\|_{0,K} \\
&\quad + C\|\varepsilon\|_{0,\infty,K} h_{y,K}^2 h_{x,K}^3 \|\partial_{xxy}p\|_{0,K} h_{x,K}^{-1} h_{y,K}^{-1} \|\partial_x v_{h1}\|_{0,K} \\
&\le C\|\varepsilon\|_{0,\infty,K} h_K^3 |p|_{3,K} |v_{h1}|_{1,K} .
\end{aligned}$$

Let K and K' denote two neighbour cells. On the common vertical edge $l_2 = l_4'$ (or $l_4 = l_2'$) we get from Hölder and inverse inequalities, Bramble-Hilbert lemma and properties of F_K

$$\begin{aligned}
&\left| \frac{1}{6}\varepsilon_K \int\limits_{l_2} \partial_{yy}p F_K^2 \partial_{yy} v_{h1} dy - \frac{1}{6}\varepsilon_{K'} \int\limits_{l_4'} \partial_{yy}p F_{K'}^2 \partial_{yy} v_{h1} dy \right| \\
&= \left| \frac{1}{6} \int\limits_{l_2} (\varepsilon_K - \varepsilon) \partial_{yy}p F_K^2 \partial_{yy} v_{h1} dy - \frac{1}{6} \int\limits_{l_4'} (\varepsilon_{K'} - \varepsilon) \partial_{yy}p F_{K'}^2 \partial_{yy} v_{h1} dy \right| \qquad (4.33) \\
&\le C \left\{ |\varepsilon_K - \varepsilon|_{0,\infty,\partial K} + |\varepsilon_{K'} - \varepsilon|_{0,\infty,\partial K'} \right\} \|\partial_{yy}p\|_{0,\partial K}\, h_K^4 \, \|\partial_{yy} v_h\|_{0,\partial K} \\
&\le C h^3 \, |\varepsilon|_{1,\infty,K\cup K'} \|p\|_{3,K} |v_h|_{1,K}
\end{aligned}$$

according to the continuity of F_K and $\partial_{yy}v_h$ at l_2 (or l_4), the trace theorem and scaling argument

$$\|r\|_{0,\partial K} \le C h^{-1/2} (\|r\|_{0,K} + h|r|_{1,K}) \qquad \forall r \in H^1(K) \qquad (4.34)$$

and due to $\varepsilon \in H^2(\Omega) \hookrightarrow C(\overline{\Omega})$. Similarly, it holds on the common horizontal edge $l_1 = l_3'$ (or $l_3 = l_1'$)

$$\begin{aligned}
&\left| \frac{4}{36} h_{x,K}^2 h_{y,K}^2 \varepsilon_K \int\limits_{l_3} \partial_{xy}p \partial_{xx} v_{h1} dx - \frac{4}{36} h_{x,K'}^2 h_{y,K'}^2 \varepsilon_{K'} \int\limits_{l_1'} \partial_{xy}p \partial_{xx} v_{h1} dx \right| \\
&\le \frac{4}{36} |\varepsilon_K| |h_{x,K}^2 h_{y,K}^2 - h_{x,K'}^2 h_{y,K'}^2| \left| \int\limits_{l_3} \partial_{xy}p \partial_{xx} v_{h1} dx \right| + \frac{4}{36} h_{x,K'}^2 h_{y,K'}^2 \left| \int\limits_{l_3} (\varepsilon - \varepsilon_K) \partial_{xy}p \partial_{xx} v_{h1} dx \right| \\
&\quad + \frac{4}{36} h_{x,K'}^2 h_{y,K'}^2 \left| \int\limits_{l_1'} (\varepsilon_K' - \varepsilon) \partial_{xy}p \partial_{xx} v_{h1} dx \right| \\
&\le C h^3 \, |\varepsilon|_{1,\infty,K\cup K'} \|p\|_{3,K} |v_{h1}|_{1,K}
\end{aligned} \qquad (4.35)$$

due to assumption (4.26). Summing up and taking into consideration that v_{h1} vanishes on the boundary $\partial\Omega$, we conclude the global bound

$$|(\bar{\varepsilon}\partial_x v_{h1}, p - j_h p)| \le C h^3 \|\varepsilon\|_{1,\infty} \|p\|_3 |v_h|_1 . \qquad (4.36)$$

The estimate

$$|(\bar{\varepsilon}\partial_y v_{h1}, p - j_h p)| \leq Ch^3 \|\varepsilon\|_{1,\infty} \|p\|_3 |v_h|_1 \tag{4.37}$$

can be obtained analogously. Collecting (4.28),(4.29),(4.30),(4.31) and (4.36) implies the assertion (4.27). Using expansion techniques from [61] one can show in analogous way the superclose estimate for $n = 3$. □

Combining above superclose estimates, yields

Theorem 4.2 *Let the weak solution $(\boldsymbol{u}, p)$ of the Stokes-like problem (2.124) satisfy the regularity assumption $(\boldsymbol{u}, p) \in \boldsymbol{H}^4(\Omega) \times H^3(\Omega)$ and let the mesh be quasi uniform in each coordinate axis. Let j_h be the L^2-projection onto M_h and $\boldsymbol{i}_h\boldsymbol{u}$ be the nonstandard interpolation onto $\boldsymbol{X}_h$ defined above. Then, for the (Q_2, P_1^{disc}) finite element solution $(\boldsymbol{u}_h, p_h)$ we have the superclose estimate*

$$|\boldsymbol{u}_h - \boldsymbol{i}_h\boldsymbol{u}|_1 + \|p_h - j_h p\|_0 \leq Ch^3 \big(|\boldsymbol{u}|_3 + |\boldsymbol{u}|_4 + \|p\|_3 \big) , \tag{4.38}$$

provided that $\varepsilon \in W^{2,\infty}(\Omega)$.

Proof. Use (4.2) and the estimates of the Lemmata 4.1 and 4.2. □

Let us assume that we have the standard error estimate

$$\|\boldsymbol{u} - \boldsymbol{u}_h\|_0 + h\left(|\boldsymbol{u} - \boldsymbol{u}_h|_1 + \|p - p_h\|_0\right) \leq C\, h^{k+1} , \tag{4.39}$$

with $C = C(Re^{-1}, \|\boldsymbol{u}\|_{k+1}, \|p\|_k)$ and a superclose property for the corresponding linear problem based on

$$|E(\boldsymbol{u}, p; \boldsymbol{v}_h, q_h)| \leq C\, h^{k+1} (|\boldsymbol{v}_h|_1 + \|q_h\|_0) , \tag{4.40}$$

where $C = C(\|\boldsymbol{u}\|_{k+2} + \|p\|_{k+1})$. Sufficient conditions that (4.39) holds true can be found in [30] for $\varepsilon = 1$. The numerical tests from Chapter 2 indicate that (4.39) holds. In Section 4.2 the validity of (4.40) for $k = 2$ has been shown in case of the $(Q_2, P_1^{\mathrm{disc}})$ element pair and appropriate interpolation operators.

Lemma 4.3 *Let the weak solution $(\boldsymbol{u}, p)$ of the Brinkman–Forchheimer problem (2.5) satisfy the regularity assumption $\boldsymbol{u} \in \boldsymbol{H}^{k+1}(\Omega)$, $k \geq 1$, and let the discrete solution $(\boldsymbol{u}_h, p_h)$ satisfy (4.39). Then,*

$$|\tilde{n}(\boldsymbol{u}, \boldsymbol{u}, \boldsymbol{v}_h) - \tilde{n}(\boldsymbol{u}_h, \boldsymbol{u}_h, \boldsymbol{v}_h)| \leq C\, h^{k+1} |\boldsymbol{v}_h|_1 \qquad \forall \boldsymbol{v}_h \in \boldsymbol{X}_{h0} \tag{4.41}$$

and

$$|d(\boldsymbol{u}; \boldsymbol{u}, \boldsymbol{v}_h) - d(\boldsymbol{u}_h; \boldsymbol{u}_h, \boldsymbol{v}_h)| \leq C\, h^{k+1} \|\beta\|_{0,\infty} |\boldsymbol{v}_h|_1 \qquad \forall \boldsymbol{v}_h \in \boldsymbol{X}_{h0} . \tag{4.42}$$

Proof. Following the proof of Lemma 4.1 from [61], we split the difference into

$$\tilde{n}(\boldsymbol{u}, \boldsymbol{u}, \boldsymbol{v}_h) - \tilde{n}(\boldsymbol{u}_h, \boldsymbol{u}_h, \boldsymbol{v}_h) =$$

$$\tilde{n}(\boldsymbol{u}-\boldsymbol{u}_h,\boldsymbol{u},\boldsymbol{v}_h)+\tilde{n}(\boldsymbol{u}_h-\boldsymbol{u},\boldsymbol{u}-\boldsymbol{u}_h,\boldsymbol{v}_h)+\tilde{n}(\boldsymbol{u},\boldsymbol{u}-\boldsymbol{u}_h,\boldsymbol{v}_h)$$

and estimate term by term. Applying Poincaré and Hölder inequalities, the fact $\|\varepsilon\|_{0,\infty}\le 1$, the continuity of the embeddings

$$\boldsymbol{H}^1(\Omega)\hookrightarrow \boldsymbol{L}^6(\Omega)\,,\quad \boldsymbol{H}^{k+1}(\Omega)\hookrightarrow \boldsymbol{W}^{1,3}(\Omega)\,,\quad \boldsymbol{H}^{k+1}(\Omega)\hookrightarrow \boldsymbol{L}^{\infty}(\Omega)\,, \tag{4.43}$$

(4.39), (2.10) and (2.45), we get for all $\boldsymbol{v}_h\in \boldsymbol{X}_{h0}$

$$\begin{aligned}
|\tilde{n}(\boldsymbol{u}-\boldsymbol{u}_h,\boldsymbol{u},\boldsymbol{v}_h)| &\le C\,\|\boldsymbol{u}-\boldsymbol{u}_h\|_0\left(|\boldsymbol{u}|_{1,3}\|\boldsymbol{v}_h\|_{0,6}+\|\boldsymbol{u}\|_{0,\infty}|\boldsymbol{v}_h|_1\right)\\
&\le C\,h^{k+1}|\boldsymbol{v}_h|_1\\
|\tilde{n}(\boldsymbol{u}_h-\boldsymbol{u},\boldsymbol{u}-\boldsymbol{u}_h,\boldsymbol{v}_h)| &\le C\,\|\boldsymbol{u}-\boldsymbol{u}_h\|_1\,\|\boldsymbol{u}-\boldsymbol{u}_h\|_1\,\|\boldsymbol{v}_h\|_1\\
&\le C\,h^{2k}|\boldsymbol{v}_h|_1\\
|\tilde{n}(\boldsymbol{u},\boldsymbol{u}-\boldsymbol{u}_h,\boldsymbol{v}_h)| &= |(\varepsilon\boldsymbol{u}\cdot\nabla\boldsymbol{v}_h,\boldsymbol{u}-\boldsymbol{u}_h)|\\
&\le C\,\|\boldsymbol{u}\|_{0,\infty}\,|\boldsymbol{v}_h|_1\,\|\boldsymbol{u}-\boldsymbol{u}_h\|_0\le C\,h^{k+1}|\boldsymbol{v}_h|_1\,.
\end{aligned}$$

Summarising all estimates we conclude (4.41).
Next, we estimate the following splitting

$$\begin{aligned}
&d(\boldsymbol{u};\boldsymbol{u},\boldsymbol{v}_h)-d(\boldsymbol{u}_h;\boldsymbol{u}_h,\boldsymbol{v}_h)\\
&=d(\boldsymbol{u};\boldsymbol{u},\boldsymbol{v}_h)-d(\boldsymbol{u}_h;\boldsymbol{u},\boldsymbol{v}_h)+d(\boldsymbol{u}_h;\boldsymbol{u},\boldsymbol{v}_h)-d(\boldsymbol{u}_h;\boldsymbol{u}_h,\boldsymbol{v}_h)\,.
\end{aligned}\tag{4.44}$$

Using the Hölder inequality, Sobolev embeddings (4.43), Poincaré inequality and (4.39), we get

$$\begin{aligned}
|d(\boldsymbol{u};\boldsymbol{u},\boldsymbol{v}_h)-d(\boldsymbol{u}_h;\boldsymbol{u},\boldsymbol{v}_h)| &\le \|\beta\|_{0,\infty}\|\boldsymbol{u}-\boldsymbol{u}_h\|_0\|\boldsymbol{u}\|_{0,3}\|\boldsymbol{v}_h\|_{0,6}\\
&\le Ch^{k+1}|\boldsymbol{v}_h|_1
\end{aligned}\tag{4.45}$$

due to $\|\beta\|_{0,\infty}\le 1.75(1-\varepsilon_0)/\varepsilon_0$. From the Hölder inequality, a priori bound (2.110), Sobolev embedding $\boldsymbol{H}^1(\Omega)\hookrightarrow\boldsymbol{L}^6(\Omega)$ follows also

$$\begin{aligned}
|d(\boldsymbol{u}_h;\boldsymbol{u},\boldsymbol{v}_h)-d(\boldsymbol{u}_h;\boldsymbol{u}_h,\boldsymbol{v}_h)| &\le \|\beta\|_{0,6}\|\boldsymbol{u}_h\|_{0,6}\|\boldsymbol{u}-\boldsymbol{u}_h\|_0\|\boldsymbol{v}_h\|_{0,6}\\
&\le Ch^{k+1}|\boldsymbol{v}_h|_1\,.
\end{aligned}\tag{4.46}$$

The assertion (4.42) follows from (4.44) by applying triangle inequality and using (4.45) and (4.46). □

As a consequence we have the superclose property for the Brinkman–Forchheimer problem:

Theorem 4.3 *Let the weak solution $(\boldsymbol{u},p)$ of the Brinkman–Forchheimer problem satisfy the regularity assumption $(\boldsymbol{u},p)\in\boldsymbol{H}^{k+2}(\Omega)\times H^{k+1}(\Omega)$, $k\ge 1$, let the discrete solution $(\boldsymbol{u}_h,p_h)$ satisfy (4.39), and let the interpolation operators $\boldsymbol{\imath}_h$, $\jmath_h$ fulfil the estimate (4.40). Then,*

$$|\boldsymbol{u}_h-\boldsymbol{\imath}_h\boldsymbol{u}|_1+\|p_h-\jmath_h p\|_0\le C(Re^{-1},\boldsymbol{u},p)\,h^{k+1}. \tag{4.47}$$

Proof. The theorem follows directly from (4.2), (4.40), and Lemma 4.3. □

Remark 4.4 *In particular, if* $\varepsilon \in W^{2,\infty}(\Omega)$ *and the weak solution* $(\boldsymbol{u}, p)$ *of the Brinkman–Forchheimer problem satisfies the regularity assumption* $(\boldsymbol{u}, p) \in \boldsymbol{H}^4(\Omega) \times H^3(\Omega)$ *we get the superclose estimate*

$$|\boldsymbol{u}_h - \boldsymbol{i}_h \boldsymbol{u}|_1 + \|p_h - j_h p\|_0 \le C(Re^{-1}, \boldsymbol{u}, p)\, h^3$$

for the (Q_2, P_1^{disc}) *element pair to the nonstandard interpolants defined in Section 4.2.*

4.3 (Q_3, P_2^{disc}) Post-processing

In this section we will define the interpolation operators $\boldsymbol{I}_{2h}$ and J_{2h} which fulfil the properties (A), (B), and (C). In the following we assume that the triangulation $\mathcal{T}_h$ was obtained from the triangulation $\mathcal{T}_{2h}$ by regular refinement, i.e., each patch $P \in \mathcal{T}_{2h}$ consists of 4 congruent child cells $K_i \in \mathcal{T}_h$, $i = 1, \dots, 4$.

In view of property (B), a suitable candidate for post-processing the discrete solution $(\boldsymbol{u}_h, p_h)$ obtained with the finite element pair (Q_2, P_1^{disc}) is the finite element pair (Q_3, P_2^{disc}), i.e., the space $\boldsymbol{Y}_{2h}$ consists of vector-valued function where each component is a continuous, piecewise Q_3 function while the space N_{2h} contains functions which are piecewise P_2 polynomials with no continuity requirements across cell borders.

To ensure property (A), the nodal functionals which define the operators $\boldsymbol{I}_{2h}$ and J_{2h} are built by using linear combinations of the nodal functionals from $\boldsymbol{i}_h$ and j_h, respectively. Note however, that arbitrary linear combinations may lead to interpolation operators which are not unisolvent.

Our aim is to construct the operators $\boldsymbol{I}_{2h}$ and J_{2h} by

$$\boldsymbol{I}_{2h}(\boldsymbol{v})|_P := \boldsymbol{I}_{2h}\left(\boldsymbol{v}|_P\right), \qquad J_{2h}(q)|_P := J_{2h}\left(q|_P\right),$$

i.e., the interpolation operators $\boldsymbol{I}_{2h}$ and J_{2h} act patch-wise locally.

The construction of the restrictions of $\boldsymbol{I}_{2h}$ and J_{2h} on P will be done via the reference patch $\widehat{P} = (-1, +1)^2$. To this end, let $\boldsymbol{F}_P : \hat{P} \to P$ be a bijective reference mapping. Then we define

$$\boldsymbol{I}_{2h}(\boldsymbol{v})|_P := \hat{\boldsymbol{I}}_{2h}(\hat{\boldsymbol{v}}_P) \circ \boldsymbol{F}_P^{-1}, \qquad J_{2h}(q)|_P := \hat{J}_{2h}(\hat{q}_P) \circ \boldsymbol{F}_P^{-1},$$

where $\hat{\boldsymbol{v}}_P = \boldsymbol{v}|_P \circ F_P$ and $\hat{q}_p = q|_P \circ \boldsymbol{F}_P$.

Since the child cells K_i are obtained by regular refinement of a patch $P \in \mathcal{T}_{2h}$, they are images of child cells $\hat{K}_i$ of $\hat{P}$.

On $\widehat{P}$ let $\widehat{X}$ denote the space of continuous, piecewise Q_2 functions and $\widehat{M}$ the space of piecewise P_1 functions.

4.3.1 Velocity post-processing

For $\hat{\boldsymbol{v}} = (\hat{v}_1, \hat{v}_2)$ we define $\hat{\boldsymbol{I}}_{2h}$ by

$$\hat{\boldsymbol{I}}_{2h}(\boldsymbol{v}) := \big(\hat{I}_{2h}(v_1), \hat{I}_{2h}(v_2)\big).$$

The scalar interpolation operator $\hat{I}_{2h}$ is defined by

$$\hat{N}_j\big(\hat{I}_{2h}(\hat{w})\big) = \hat{N}_j(\hat{w}), \qquad j = 1, \dots, 16,$$

where the nodal functionals $\hat{N}_j$, $j = 1, \dots, 16$, are given by

$$\begin{aligned}
\hat{N}_j(\hat{w}) &:= \hat{w}(\hat{\boldsymbol{a}}_j), & j &= 1, \dots, 4,\\
\hat{N}_{j+8}(\hat{w}) &:= \frac{1}{|\hat{l}_j|} \int\limits_{\hat{l}_j} \hat{w}\, d\hat{s}, & j &= 1, \dots, 8,\\
\hat{N}_{i+12}(\hat{w}) &:= \frac{1}{|\widehat{K}_i|} \int\limits_{\widehat{K}_i} \hat{w}\, d\xi\, d\eta\,, & i &= 1, \dots, 4.
\end{aligned}$$

Here, $\hat{\boldsymbol{a}}_j$, $j = 1, \dots, 4$, are the vertices of $\widehat{P}$; $\hat{l}_j$, $j = 1, \dots, 8$, are all outer edges of children of $\widehat{P}$, i.e., $\hat{l}_j \subset \partial\widehat{P}$; and $\widehat{K}_i$, $i = 1, \dots, 4$, are the children of $\widehat{P}$, see Figure 4.2.

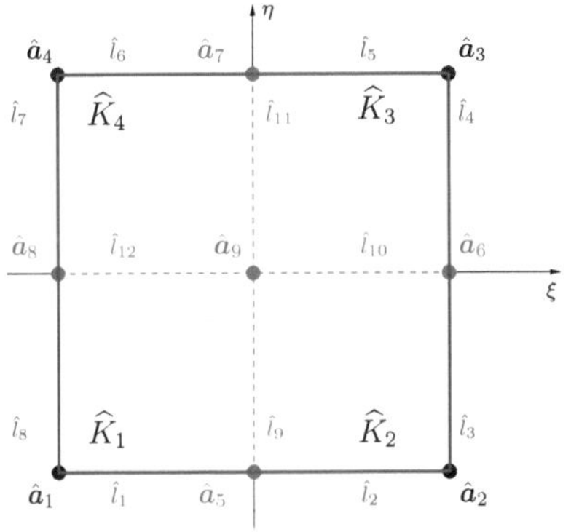

Figure 4.2: Macro-cell $\widehat{P} = (-1, 1)^2$

Note that due to the embedding $H^2(\Omega) \hookrightarrow C^0(\overline{\Omega})$ all nodal functionals are well defined and that the locally on each patch P defined interpolations $\boldsymbol{I}_{2h}$ build a continuous global interpolation, which we denote again by $\boldsymbol{I}_{2h}$.

Since the nodal functionals $\hat{N}_j$, $j = 1, \ldots, 16$, are suitable linear combinations of nodal functionals for $\hat{i}_h$ the property (A) is fulfilled. It is easy to check that $\hat{I}_{2h}$ is Q_3-unisolvent and any Q_3 function on $\widehat{P}$ is reproduced. Thus, the property (B) follows from the Bramble–Hilbert lemma together with the standard estimates for the reference mapping $\boldsymbol{F}_P$. Since $|\hat{I}_{2h} \cdot|_{1,\widehat{P}}$ and $|\cdot|_{1,\widehat{P}}$ are norms on the finite dimensional space $\widehat{X}/\mathbb{R}$ we have the equivalence

$$|\hat{I}_{2h}(\hat{v}_i)|_{1,\widehat{P}} \leq C|\hat{v}_i|_{1,\widehat{P}} \qquad \forall v_i \in \widehat{X},\ i = 1, 2\,.$$

Using the estimates for $\boldsymbol{F}_P$ we obtain

$$|\boldsymbol{I}_{2h}(\boldsymbol{v}_h)|_{1,P} \leq C|\boldsymbol{v}_h|_{1,P} \qquad \forall \boldsymbol{v}_h \in \boldsymbol{X}_h$$

with a constant C independent of P. Property (C) follows by summing up over all patches P.

4.3.2 Pressure post-processing

We define the following unions of child cells $\hat{K}_i$, see Figure 4.3. Let $\hat{D}_j$, $j = 1, \ldots, 2$, denote the union of each two diagonally lying children. Let $\hat{A}_\xi^1$ and $\hat{A}_\xi^2$ be the union of those children which lie in the half space $\xi < 0$ and $\xi > 0$, respectively. The unions $\hat{A}_\eta^1$, $\hat{A}_\eta^2$ are defined in an analogous way.

We define the interpolation operator $\hat{J}_{2h}$ according to

$$\hat{M}_i(\hat{J}_{2h}(\hat{q})) = \hat{M}_i(\hat{q}), \qquad i = 1, \ldots, 6\,,$$

where

$$\hat{M}_j(\hat{q}) := \frac{1}{|\widehat{D}_j|} \int_{\widehat{D}_j} 1 \cdot \hat{q}\, d\xi\, d\eta, \qquad j = 1, 2,$$

$$\hat{M}_{j+2}(\hat{q}) := \frac{1}{|\widehat{A}_\xi^j|} \int_{\widehat{A}_\xi^j} \xi \cdot \hat{q}\, d\xi\, d\eta, \qquad j = 1, 2,$$

$$\hat{M}_{j+4}(\hat{q}) := \frac{1}{|\widehat{A}_\eta^j|} \int_{\widehat{A}_\eta^j} \eta \cdot \hat{q}\, d\xi\, d\eta \qquad j = 1, 2.$$

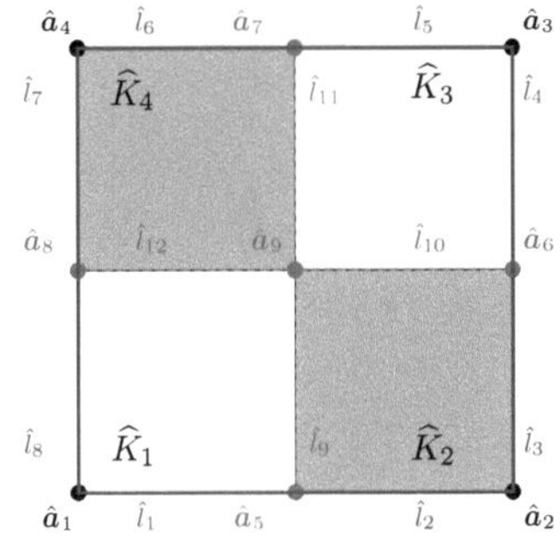

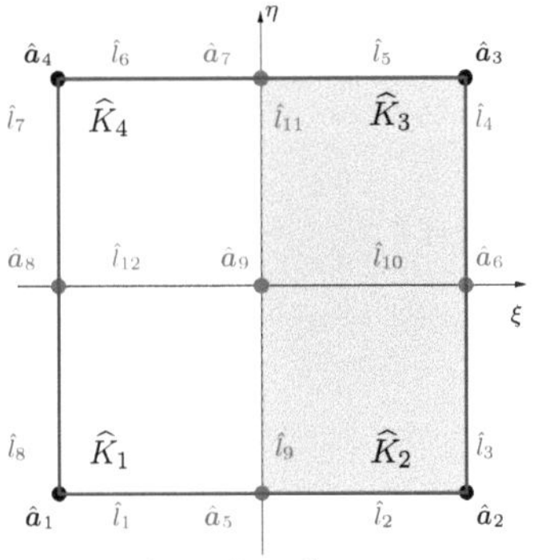

Figure 4.3: Examples of union cells required for pressure-postprocessing

One can check that these nodal functionals are P_2-unisolvent and that the interpolation operator $\hat{J}_{2h}$ reproduces P_2 polynomials on $\widehat{P}$. Together with the estimates for the reference mapping $\boldsymbol{F}_P$ and the Bramble–Hilbert lemma this results in property (B). Furthermore, the above given nodal functionals are linear combinations of the nodal functionals used for defining $\hat{j}_h$. Thus, property (A) is fulfilled. For the proof of property (C) we use the fact that $\hat{J}_{2h}$ is a linear and unisolvent interpolation operator. Furthermore, we apply again estimates for the reference mapping and its inverse and we exploit the equivalence of the norm $\|\hat{J}_{2h}\cdot\|_{0,\widehat{K}}$ and $\|\cdot\|_{0,\widehat{K}}$ on the finite dimensional space $\widehat{M}/\mathbb{R}$.

Remark 4.5 *Due to the construction of J_{2h} we get that $p_h \in L_0^2(\Omega)$ implies $J_{2h}(p_h) \in L_0^2(\Omega)$.*

4.3.3 Superconvergence result

Now we can formulate our superconvergence result.

Theorem 4.6 *Let $(\boldsymbol{u}, p)$ be the solution of the Stokes-like or Brinkman–Forchheimer problem which fulfils the regularity assumption $\boldsymbol{u} \in \boldsymbol{H}^4(\Omega)$ and $p \in H^3(\Omega)$. Furthermore, let $(\boldsymbol{u}_h, p_h)$ be the discrete solution of the corresponding problem with the finite element pair (Q_2, P_1^{disc}). Then, we have the estimate*

$$|\boldsymbol{u} - \boldsymbol{I}_{2h}\boldsymbol{u}_h|_1 + \|p - J_{2h}p_h\|_0 \le C(Re^{-1}, \boldsymbol{u}, p)\, h^3\,, \tag{4.48}$$

provided that $\varepsilon \in W^{2,\infty}(\Omega)$.

Proof. The proof follows directly from the superclose estimates (4.38) and (4.47), respectively, and the properties (A), (B), and (C). □

Remark 4.7 *In order to post-process the discrete solution* $(\boldsymbol{u}_h, p_h) \in \boldsymbol{X}_h \times M_h$ *we need only the action of the locally defined operators* $\boldsymbol{I}_{2h}$ *and* J_{2h} *on discrete functions. These mappings can be represented by local matrices which describe the action of* $\boldsymbol{I}_{2h}$ *and* J_{2h} *on* $\boldsymbol{v}_h$ *and* q_h*, respectively, in terms of appropriate bases.*

4.4 Numerical results

The test problem from Chapter 2 with the smooth solution (2.125) was computed on a family of uniform meshes. Level 0 corresponds to a partition of the unit square $\Omega = (0,1)^2$ into 4 subcubes. A refinement step divides each mesh cell into 4 congruent cells.

Table 4.1: Velocity errors in the Brinkman–Forchheimer problem, $Re = 1.0$, discretised with (Q_2, P_1^{disc}), post-processed by (Q_3, P_2^{disc}).

l.	$\|\boldsymbol{u} - \boldsymbol{u}_h\|_1$	order	$\|\boldsymbol{u}_h - \boldsymbol{i}_h\boldsymbol{u}\|_1$	order	$\|\boldsymbol{u} - \boldsymbol{I}_h\boldsymbol{u}_{h/2}\|_1$	order
0	1.114e+0		5.157e+0		2.493e−1	
1	2.799e−1	1.992	6.086e−2	3.083	3.015e−2	3.048
2	6.531e−2	2.100	5.463e−3	3.478	5.590e−3	2.431
3	1.642e−2	1.991	6.334e−4	3.108	7.097e−4	2.978
4	4.113e−3	1.998	7.838e−5	3.015	8.901e−5	2.995
5	1.029e−3	1.999	9.808e−6	2.998	1.114e−5	2.999
6	2.572e−4	2.000	1.229e−6	2.997		

The errors for the Brinkman-Forchheimer problem (with $Re = 1.0$) are shown in Tables 4.1 and 4.2. Note that in order to determine the post-processed (Q_3, P_2^{disc}) solution $(\boldsymbol{I}_h\boldsymbol{u}_{h/2}, J_h p_{h/2})$ on the level l the discrete (Q_2, P_1^{disc}) solution $(\boldsymbol{u}_{h/2}, p_{h/2})$ on the next finer level $l+1$ has to be calculated. The convergence rates are in good agreement with the theoretical rates given in Theorem 4.6. This can also be seen in the Figure 4.4. The benefit of the post-processing is visible in this figure. In order to achieve the accuracy of the (Q_2, P_1^{disc}) solution on level $l = 5$ it is sufficient to determine the post-processed (Q_3, P_2^{disc}) solution on 2-3 level coarser. In Table 4.3 we compare the errors of two third order methods, namely the post-processed (Q_2, P_1^{disc}) solution and the (Q_3, P_2^{disc}) solution. The two discrete solutions are compared on the same mesh, which means that the lower order solution has been determined on the next finer mesh. Since the errors are approximately of the same size and the number of degrees of freedom for the (Q_2, P_1^{disc}) solution on the next finer level is higher compared to those for the (Q_3, P_2^{disc}) solution, it seems

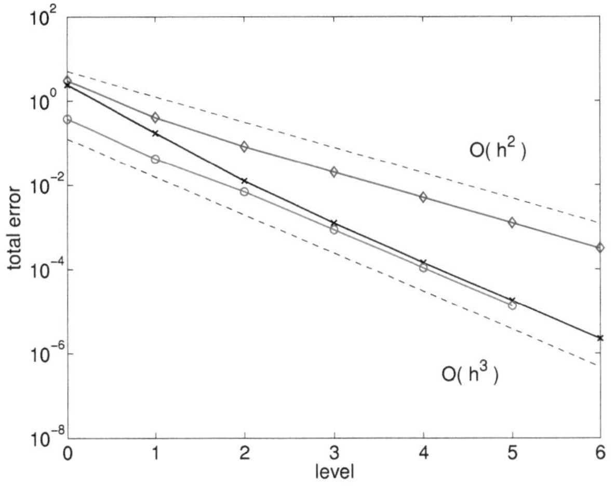

Figure 4.4: Convergence rates for the Brinkman–Forchheimer problem, $Re = 1.0$.
$-\diamond-$ (Q_2, P_1^{disc}) finite element error $|\boldsymbol{u} - \boldsymbol{u}_h|_1 + \|p - p_h\|_0$,
$-\times-$ superclose error $|\boldsymbol{u}_h - \boldsymbol{i}_h\boldsymbol{u}|_1 + \|p_h - j_h p\|_0$,
$-\circ-$ post-processing error $|\boldsymbol{u} - \boldsymbol{I}_h\boldsymbol{u}_{h/2}|_1 + \|p - J_h p_{h/2}\|_0$.

Table 4.2: Pressure errors in the Brinkman–Forchheimer problem, $Re = 1.0$, discretised with (Q_2, P_1^{disc}), post-processed by (Q_3, P_2^{disc}).

l.	$\|p - p_h\|_0$	order	$\|p_h - j_h p_h\|_0$	order	$\|p - J_h p_{h/2}\|_0$	order
0	1.886e−0		1.873e+0		1.286e−1	
1	1.257e−1	3.907	1.108e−1	4.079	1.183e−2	3.443
2	1.663e−2	2.918	6.941e−3	3.997	1.351e−3	3.130
3	3.845e−3	2.113	6.147e−4	3.497	1.641e−4	3.041
4	9.523e−4	2.013	6.684e−5	3.201	2.040e−5	3.008
5	2.377e−4	2.002	8.019e−6	3.059	2.547e−6	3.002
6	5.940e−5	2.001	9.926e−7	3.014		

to be inefficient to use the post-processing technique. But the opposite is the case: when solving the discrete systems by a multilevel approach the costs to solve the discrete system for the (Q_3, P_2^{disc}) solution is much higher than those for the (Q_2, P_1^{disc}) solution on the next finer mesh, which means that the size of the system (number of degrees of freedom) is not an appropriate measure for the solving complexity. Indeed, if different discretisations

Table 4.3: Errors of the post-processed (Q_2, P_1^{disc}) solution $(\boldsymbol{I}_h\boldsymbol{u}_{h/2}, J_h p_{h/2})$ and the (Q_3, P_2^{disc}) finite element solution $(\boldsymbol{w}_h, r_h)$ for the Brinkman–Forchheimer problem, $Re = 1.0$.

l.	$\|\boldsymbol{u} - \boldsymbol{I}_h\boldsymbol{u}_{h/2}\|_1$	$\|\boldsymbol{u} - \boldsymbol{w}_h\|_1$	$\\|p - J_h p_{h/2}\\|_0$	$\\|p - r_h\\|_0$
0	2.493e−1	2.512e−1	1.286e−1	8.567e−2
1	3.015e−2	3.047e−2	1.183e−2	8.329e−3
2	5.590e−3	5.638e−3	1.351e−3	1.005e−3
3	7.097e−4	7.160e−4	1.641e−4	1.221e−4
4	8.901e−5	8.983e−5	2.040e−5	1.511e−5
5	1.114e−5	1.124e−5	2.547e−6	1.882e−6

are compared, the costs for generating and solving the local systems in the smoothing procedure are more important. Thus, although there are more unknowns to be determined for the (Q_2, P_1^{disc}) solution on the next finer level, we can solve the system cheaper and obtain by post-processing the same accuracy with less overall effort.

5 Physically reliable stabilisation method for scalar problems

Convection-diffusion-reaction equations occur for instance in chemical engineering. Depending on the problem, different types of boundary conditions are applied on different parts of the domain boundary. A common feature of these problems is the small diffusion coefficient, i.e., the process is convection and/or reaction dominant. Since standard Galerkin discretisations will produce unphysical oscillations for this type of problems, stabilisation techniques have been developed. The streamline-upwind Petrov–Galerkin method (SUPG) has been successfully applied to convection-diffusion-reaction problems. It was proposed by Hughes and Brooks [40]. One fundamental drawback of SUPG is that several terms which include second order derivatives have to be added to the standard Galerkin discretisation in order to ensure consistency. Alternatively, continuous interior penalty methods [2, 15], residual free bubble method [27, 28, 29], or subgrid modelling [23, 32] can be used for stabilising the discretised convection-diffusion-reaction problems.

Despite of well investigated stabilising effects of the local projection method (LPS) for scalar convection-diffusion problems and its relations to other stabilisation methods like SUPG and continuous interior penalty methods (CIP), see [63, 69, 51], the problem of spurious oscillations at the boundary layer arises. This lack of numerical stability can lead to solutions which do not preserve physical properties, e.g., non-negativity of concentration of chemical species. Expressing this issue mathematically, we can say that numerical solutions do not satisfy the maximum principle in the certain sense. The pioneering work on the field of discrete maximum principle for finite elements is [18]. Authors established there that the solution with continuous piecewise linears satisfies the discrete maximum principle for Poisson equation on weakly acute triangular meshes. Since then many many improvements and extensions have been done. We mention [43, 81, 21, 79, 78, 77, 73, 50, 14, 49, 34, 48, 52] as the most important results of the last decades. Undesired spurious oscillations can be also reduced or even eliminated by employing a suitable choice of stabilising parameters in order to get a nodally exact solutions, see [69]. Another possibility of satisfying the discrete maximum principle is the use of additional terms, see review article [44] and [45] for the detailed discussion of the optimal choice of stabilising parameters. It has been proved that the first-order artificial viscosity scheme of [20] and nonlinear artificial scheme of [16] produce solutions which satisfy the discrete maximum principle. The rigorous proof for discrete maximum principle for CIP scheme perturbated by shock capturing term has been established in [17]. Since our proof techniques do not involve algebraic arguments and the behaviour of the discretised Laplace operator is well studied for triangular meshes, we decide to change our mesh topology into triangles.

Our aim is to establish a discrete maximum principle of the local projection scheme modified by shock capturing (LPSSC) and to provide the convergence theory of this method applied to convection-diffusion-reaction problems with Dirichlet boundary conditions. Furthermore, several test problems with different types of interior and boundary layers will be presented. They show that the local projection stabilisation with the edge oriented shock capturing allows to obtain numerical solutions which are physically reliable.

5.1 Model problem and local projection method with shock capturing

5.1.1 Weak formulation

We consider the following Dirichlet problem for the scalar convection-diffusion-reaction equation in two dimensions

$$\left.\begin{aligned} -D\Delta u + \boldsymbol{b}\cdot\nabla u + cu = f \quad & \text{in } \Omega, \\ u = g \quad & \text{on } \Gamma = \partial\Omega\,, \end{aligned}\right\} \tag{5.1}$$

where $D > 0$ is a small diffusion constant. We are looking for the distribution of concentration or temperature u in a polygonally bounded reactor domain $\Omega \subset \mathbb{R}^2$. The reaction coefficient $c \in L^\infty(\Omega)$ is assumed to be non-negative. Let $f \in L^2(\Omega)$, $g \in H^{1/2}(\Gamma)$ be given functions. Furthermore, we require that the convection field $\boldsymbol{b} \in \left(W^{1,\infty}(\Omega)\right)^n$, $n = 2$, and the reaction coefficient c fulfils for some $c_0 > 0$ the following condition

$$c(x) - \frac{1}{2}\nabla\cdot\boldsymbol{b}(x) \geq c_0 > 0 \qquad \forall x \in \overline{\Omega}. \tag{5.2}$$

We define the function spaces

$$V = H^1(\Omega) \quad \text{and} \quad V_0 = \{v \in V :\, v|_\Gamma = 0\}.$$

A weak formulation of (5.1) reads

Find $u \in V$ with $u|_\Gamma = g$ such that

$$a(u,v) = (f,v) \quad \forall\, v \in V_0 \tag{5.3}$$

where the bilinear form $a : H^1(\Omega)\times H^1(\Omega) \to \mathbb{R}$ is defined by

$$a(u,v) = D(\nabla u, \nabla v) + (\boldsymbol{b}\cdot\nabla u, v) + (cu, v). \tag{5.4}$$

The condition (5.2) guarantees the V_0-coercivity of the bilinear form $a(\cdot,\cdot)$. The existence and uniqueness of a weak solution of problem (5.3) can be concluded from the Lax–Milgram lemma. In the following we assume that the weak solution is H^2 regular. The higher regularity of the weak solution of Dirichlet problems on convex domains is a well known fact, see [31].

5.1.2 Stabilisation by local projection

For the finite element discretisation of (5.3), we are given a shape regular family $\{\mathcal{T}_h\}$ of decomposition of Ω into triangles. The diameter of the triangular cell K will be denoted by h_K and the mesh size parameter h is defined by $h := \max_{K\in\mathcal{T}_h} h_K$. For $\mathcal{T}_h$, let $\mathcal{E}_h$ denote the set of all interior edges E of cells $K \in \mathcal{T}_h$ which belong to Ω.

It is a well known fact that the standard Galerkin discretisation of (5.3) can fail in the convection dominated regime $D \ll 1$. In the following, we use stabilisation method by local projection.

Let $\widehat{K}$ denote a reference cell and $\boldsymbol{F}_K : \widehat{K} \to K$ the affine mapping from the reference triangle onto an arbitrary cell $K \in \mathcal{T}_h$. Furthermore, let

$$\hat{b}(\hat{x},\hat{y}) = 27\prod_{i=1}^{3}\hat{\lambda}_i(\hat{x},\hat{y})$$

be the normalised bubble function which achieves the value 1 at the barycentre of $\widehat{K}$. In the above definition $\hat{\lambda}_i$, $i = 1,2,3$, are barycentric coordinates on $\widehat{K}$. The local linear space enriched by the cubic bubble $\hat{b}$ is denoted by

$$P_1^+(\widehat{K}) := P_1(\widehat{K}) \oplus \operatorname{span}\{\hat{b}\}\,.$$

Let

$$(V_h, D_h) := (P_{1,h}^+, P_{0,h}^{\text{disc}}) \tag{5.5}$$

be the pair of finite element spaces defined *via* the reference mapping

$$\begin{aligned} P_{1,h}^+ &:= \{v \in H^1(\Omega) \,:\, v|_K \circ \boldsymbol{F}_K \in P_1^+(\widehat{K}) \quad \forall K \in \mathcal{T}_h\}\,,\\ P_{0,h}^{\text{disc}} &:= \{v \in L^2(\Omega) \,:\, v|_K \circ \boldsymbol{F}_K \in P_0(\widehat{K}) \quad \forall K \in \mathcal{T}_h\}\,. \end{aligned}$$

Furthermore, we introduce the linear finite element space

$$P_{1,h} := \{v \in H^1(\Omega) \,:\, v|_K \circ \boldsymbol{F}_K \in P_1(\widehat{K}) \quad \forall K \in \mathcal{T}_h\}\,.$$

and set

$$V_h^L := P_{1,h}\,.$$

The following approximation property of fluctuation operator can be stated

$$\|\kappa_h q\|_{0,K} \leq Ch|q|_{1,K} \quad \forall\, q \in H^1(K)\,,\ \forall\, K \in \mathcal{T}_h\,. \tag{5.6}$$

Moreover, the pair of enriched continuous piecewise linears and discontinuous piecewise constants fulfils the compatibility condition (3.12), see [62, Lemma 4.1]. Then, the existence of special interpolant $j_h : H^2(\Omega) \to V_h$ satisfying the orthogonality condition (3.14) and approximation properties (3.15) is provided according to Theorem 3.1.
We define the stabilising term $S_h : V \times V \to \mathbb{R}$ in the usual way

$$S_h(u_h, v_h) := \sum_{K\in\mathcal{T}_h} \tau_K\big(\kappa_h(\nabla u_h), \kappa_h(\nabla v_h)\big)_K \tag{5.7}$$

where $\tau_K = \tau_0 h_K$, $\tau_0 > 0$, $K \in \mathcal{T}_h$, denote user-defined parameters. Let

$$V_{h0} = \{v \in V_h :\ v_h|_\Gamma = 0\}$$

be the discrete test space and

$$V_{h0}^L = \{v \in V_h^L :\ v_h|_\Gamma = 0\}$$

its part containing piecewise linears. The local projection stabilisation of the discretisation of (5.3) reads as follows

Find $u_h \in V_h$ with $u_h|_\Gamma = g_h$ such that

$$a(u_h, v_h) + S_h(u_h, v_h) = (f, v_h) \quad \forall\, v_h \in V_{h0} \tag{5.8}$$

where g_h denotes a suitable approximation of g_h, e.g. $g_h = j_h g$. For the associated local projection norm

$$|||v_h||| := \left\{D|v_h|_1^2 + c_0\|v_h\|_0^2 + S_h(v_h, v_h)\right\}^{1/2} \tag{5.9}$$

the following error estimate holds

$$|||u - u_h||| \leq C\left(\sum_{K\in\mathcal{T}_h} (D + h_K)h_K^2\, \|u\|_{2,K}^2\right)^{1/2}, \tag{5.10}$$

see [63].

5.1.3 Edge oriented shock capturing scheme

The numerical examples given in [63] lead us to suspect that the discrete solution exhibits in general undesired spurious oscillations at boundary layers. The effect of non-uniform pointwise convergence is also known for discrete solutions obtained by SUPG method and in general can be explained as manifestation of Gibb's phenomenon, see [23]. To circumvent this shortcoming we propose the perturbation of the stabilised bilinear form $a(\cdot,\cdot)+S_h(\cdot,\cdot)$ by the semilinear shock capturing operator $j_h : V_h \times V_h \to \mathbb{R}$ which adds a lot of dissipation in the regions where the discrete solution is oscillatory and does not act in the regions where the solution is constant. Let us decompose the finite element solution $u_h \in V_h$ in the following way

$$u_h = u_h^L + u_h^B\,, \tag{5.11}$$

where $u_h^L \in V_h^L$ and $u_h^B \in V_h \setminus V_h^L$ denote linear and bubble part of u_h, respectively. In the spirit of [17] we define the edge oriented shock capturing operator

$$j_h(u_h; v_h) := c_p \sum_{E\in\mathcal{E}_h} \gamma_E(\boldsymbol{b}, c)\, h_E^3 \operatorname{sign}\left(\frac{\partial u_h}{\partial t_E}\right) \left|[\![\nabla u_h^L]\!]_E\right| \frac{\partial v_h}{\partial t_E}\,, \tag{5.12}$$

where $c_p > 1$ is a user-defined parameter, the constant $\gamma_E(\boldsymbol{b}, c) > 0$ will be exploited in the next section, h_E denotes the length of the edge E, and $\dfrac{\partial v_h}{\partial t_E}$ stands for the tangential derivative of v_h. We define by

$$[\![\nabla u_h]\!]_E := \left((\nabla u_h)|_K \cdot \boldsymbol{n}_K + (\nabla u_h)|_{K'} \cdot \boldsymbol{n}_{K'}\right)\Big|_E\,.$$

the scalar jump of ∇u_h across the edge $E = K \cap K'$. For $x \in \mathbb{R}$ we set $\operatorname{sign}(x) := \frac{x}{|x|}$ for $x \neq 0$ and $\operatorname{sign}(0) := 0$. We note that the operator j is nonlinear in the first argument and it holds

$$j_h(-u_h; v_h) = -j_h(u_h; v_h)\,. \tag{5.13}$$

Our discrete shock capturing scheme reads as follows

Find $u_h \in V_h$ with $u_h|_\Gamma = g_h$ such that

$$a(u_h, v_h) + S_h(u_h, v_h) + j_h(u_h; v_h) = (f, v_h) \quad \forall\, v_h \in V_{h0}\,. \tag{5.14}$$

In the next section we want to answer the question whether the scheme (5.14) with the edge oriented shock capturing operator obeys the discrete maximum principle.

5.2 Discrete maximum principle

5.2.1 Maximum principle for continuous problem

First, we recall results concerning maximum principle for the classical solutions of continuous problem (5.1). Following the textbook of Evans [25], we state

Theorem 5.1 *Let $u \in C^2(\Omega) \cap C(\overline{\Omega})$ be a classical solution of (5.1). Then, it holds*

$$f \le 0 \quad in \quad \Omega \qquad \Rightarrow \qquad \max_{x\in\overline{\Omega}} u(x) \le \max\{0, \max_{x\in\partial\Omega} u(x)\} \tag{5.15}$$

and

$$f \geq 0 \quad in \quad \Omega \qquad \Rightarrow \qquad \min_{x \in \overline{\Omega}} u(x) \geq \min\{0, \min_{x \in \partial\Omega} u(x)\}\,. \tag{5.16}$$

Theorem 5.1 tells us nothing else than that the solution achieves extrema on the boundary, provided that the right hand side f and g are properly signed. From the numerics of elliptic partial differential equations we know already that many properties of the elliptic operators are transferred to their discrete counterparts. The natural question which now arises is whether the solution of (5.14) can satisfy the maximum principle in the certain sense.

5.2.2 Discrete maximum principle (DMP)

Let P_i, $i = 1, \ldots, N$ and P_i, $i = N+1, \ldots, M$ denote interior and boundary vertices resulting from the triangulation $\mathcal{T}_h$, respectively. We confine ourselves to study the principle of discrete minimum. By analogy to (5.16) we want to show that the discrete solution of (5.14) satisfies

$$f \geq 0 \quad \text{in} \quad \Omega \qquad \Rightarrow \qquad \min_{i=1,\ldots,N+M} u_h(P_i) \geq \min\{0, \min_{i=N+1,\ldots M} u_h(P_i)\}\,. \tag{5.17}$$

Let us assume that the following abstract semilinear scheme is solvable.

Find $u_h \in V_h$ with $u_h|_\Gamma = g_h$ such that

$$A(u_h; v_h) = (f, v_h) \quad \forall\, v_h \in V_{h0}\,. \tag{5.18}$$

Hereby, $A(\cdot;\cdot) : V_h \times V_h \to \mathbb{R}$ is linear in the second argument.

We denote by Ω_i, for $i = 1, \ldots, N$, the union of all cells $K \in \mathcal{T}_h$ which share the vertex P_i. Furthermore, let $\boldsymbol{n}_K$ be the outer unit normal on ∂K, and let $\mathcal{E}(P_i)$ denote a set of all edges E to which a vertex P_i belongs. We define the set of all indices j of vertices p_j that are neighbours of p_i by $\Lambda_i := \{j \neq i : \;\; \exists\, K \in \mathcal{T}_h \;\text{ with }\; p_i,\, p_j \in K\}$.

Now, we define the discrete minimum principle property of schemes with a semilinear form $A(\cdot\,;\cdot)$.

Definition 5.2 *A semilinear form $A(\cdot\,;\cdot)$ satisfies discrete minimum principle property (DMPP) if for all $u_h \in V_h$, $u_h|_\Gamma = g_h$ and for all interior vertices P_i, $i = 1, \ldots, N$ the following holds:*
If u_h takes at P_i a negative local minimum over the corresponding patch Ω_i, then there exist $\alpha_E > 0$ such that

$$A(u_h; \varphi_i) \leq -\sum_{E \in \mathcal{E}(P_i)} \alpha_E \left| [\![\nabla u_h^L]\!]_E \right|\,, \tag{5.19}$$

where $\varphi_i \in P_1$ denotes a nodal Lagrange base function associated with P_i, i.e., $\varphi_i(P_j) = \delta_{ij}$ for all $j \in \{1, \ldots, N+M\}$.

The following theorem can be applied to the general framework of shock capturing schemes.

Theorem 5.3 *Let a semilinear form $A(\cdot;\cdot)$ satisfy DMP of Definition 5.2 and $(f,\varphi_i) \geq 0$. Then, for the finite element solution of (5.18) it holds:*

$$u_h(P_i) \geq \min_{j=N+1,\dots,M} \{0, u_h(P_j)\} \quad \forall\, i = 1,\dots,N+M\,. \tag{5.20}$$

Proof. We observe, that if u_h attains a nonnegative minimum, or a negative minimum at a boundary vertex, then the assertions follows immediately. Let u_h take a global negative minimum at P_i, $i \in \{1,\dots,N\}$. Due to assumptions we get from (5.18)

$$0 \leq (f,\varphi_i) = A(u_h;\varphi_i) \leq -\sum_{E\in\mathcal{E}(P_i)} \alpha_E \,|\, [\![\nabla u_h^L]\!]_E \,|$$

which implies that u_h^L is already constant over $\overline{\Omega}_i$ and the minimum is attained also at a boundary node of Ω_i. Next, we proceed the same on the patch containing the boundary vertex of Ω_i. Repeating this argument we reach some boundary node of $\partial\Omega$. □

Remark 5.4 *The relation (5.17) with the changed sign holds for the discrete maximum property.*

Now, we prove DMP for the semilinear form

$$A(u_h;v_h) := a(u_h,v_h) + S_h(u_h,v_h) + j_h(u_h;v_h)$$

by estimating all appearing terms separately. We start with the estimate of the Laplacian.

Lemma 5.5 *Let φ_i be the piecewise linear Lagrange basis function that satisfies $\varphi_i(P_j) = \delta_{ij}$. If u_h achieves a local minimum at an inner vertex P_i, i.e.,*

$$u_h(P_j) \geq u_h(P_i) \qquad \forall\, j \in \Lambda_i\,,$$

then

$$(\nabla u_h, \nabla\varphi_i) \leq 0 \tag{5.21}$$

Proof. Let ϕ_K^B denote the bubble function on K. Then, the bubble part of the solution $u_h|_K$ can be represented as $u_h^B|_K = u_K\varphi_K^B$, $u_K \in \mathbb{R}$. First, we observe

$$(\nabla u_h^B, \nabla\varphi_i) = (\nabla u_h^B, \nabla\varphi_i)_{\Omega_i} = \sum_{K\subset\Omega_i} u_K(\nabla\varphi_K^B, \nabla\varphi_i)_K = \sum_{K\subset\Omega_i} u_K \int_{\partial K} \varphi_K^B \frac{\partial\phi_i}{\partial \boldsymbol{n}}\, ds = 0 \tag{5.22}$$

due to the integration by parts and $\varphi_K^B|_{\partial K} = 0$. Then, following [69], we get for $u_h = u_h^L + u_h^B$

$$(\nabla u_h, \nabla \varphi_i) = (\nabla u_h^L, \nabla \varphi_i) = \sum_{E\in\mathcal{E}_h(P_i)} \frac{h_E}{2} [\![\nabla u_h^L]\!]_E \,,$$

and it holds for an inner vertex P_i

$$(\nabla u_h, \nabla \varphi_i) = - \sum_{E\in\mathcal{E}_h(P_i)} \frac{h_E}{2} \left|[\![\nabla u_h^L]\!]_E\right| \,,$$

provided that u_h has at P_i a local minimum. □

Now, we estimate $\left|(\nabla u_h^L)|_K\right|$ by the certain sum of jumps $\left|[\![\nabla u_h^L]\!]_E\right|$.

Corollary 5.6 *If $w_h \in P_{1,h}$ with $w_h|_\Gamma = g_h$, has a local minimum at the vertex P_i, then*

$$\left|(\nabla w_h)|_K\right| \leq \sum_{E\in\mathcal{E}_h(P_i)} \left|[\![\nabla w_h]\!]_E\right| \qquad \forall\, K \subset \Omega_i \,. \tag{5.23}$$

Proof. See [17, Lemma 2.7]. □

In the following, we assume that the family of meshes $\{\mathcal{T}_h\}_{h>0}$ is shape regular. This implies that there exists a positive constant ρ such that

$$\max_{K\subset\Omega_i} |K| \leq \rho \min_{E\in\mathcal{E}(P_i)} h_E^2 \,, \tag{5.24}$$

and that the maximum number of cells contained in Ω_i is bounded independently of the mesh size h. Then, we can estimate the convection and reaction term.

Lemma 5.7 *Let $f = 0$. If u_h takes a local minimum at an inner vertex P_i, then*

$$(\boldsymbol{b}\cdot\nabla u_h, \nabla\varphi_i) + (cu_h, \varphi_i) + S_h(u_h, \varphi_i) \leq \sum_{E\in\mathcal{E}_h(P_i)} \gamma_{1,E}(\boldsymbol{b}, c)\frac{h_E}{2} \left|[\![\nabla u_h^L]\!]_E\right| \,, \tag{5.25}$$

whereby φ_i is the piecewise linear Lagrange basis function that satisfies $\varphi_i(P_j) = \delta_{ij}$ and the positive quantities $\gamma_{1,E}(\boldsymbol{b}, c)$ depend on the data $\boldsymbol{b}$ and c.

Proof. Let u_h take a negative local minimum at the inner vertex P_i. Using Corollary 5.6, we get for the piecewise linear Lagrange basis function φ_i with $\varphi_i(P_j) = \delta_{ij}$ and for the family of shape regular meshes $\{\mathcal{T}_h\}_{h>0}$, the following estimate

$$\begin{aligned}
(\boldsymbol{b}\cdot\nabla u_h^L, \varphi_i) &= \sum_{K\in\Omega_i} (\boldsymbol{b}\cdot\nabla u_h^L, \varphi_i)_K \leq \sum_{K\subset\Omega_i} \frac{|K|}{3} \|\boldsymbol{b}\|_{0,\infty,K} \left|(\nabla u_h^L)|_K\right| \\
&\leq \sum_{E\in\mathcal{E}(P_i)} h_E^2 \left|[\![\nabla u_h^L]\!]_E\right| \left(\sum_{K\subset\Omega_i} \frac{|K|}{3h_E^2} \|\boldsymbol{b}\|_{0,\infty,\Omega_i} \right) \\
&\leq C_1 \sum_{E\in\mathcal{E}(P_i)} h_E^2 \|\boldsymbol{b}\|_{0,\infty,\omega(E)} \left|[\![\nabla u_h^L]\!]_E\right| \,,
\end{aligned} \tag{5.26}$$

whereby we set $\omega(E) = \Omega_i \cup \Omega_j$ if the edge E joins the vertices P_i and P_j. Now, we consider the reaction term for $K \subset \Omega_i$. If u_h^L changes the sign in K, then $\|u_h^L\|_{0,\infty,K} \le h_K \|\nabla u_h\|_{0,\infty,K}$ and we get

$$(cu_h^L, \varphi_i)_K \le \|c\|_{0,\infty,K} \|u_h^L\|_{0,\infty,K} \|\varphi_i\|_{0,1,K} \le \frac{|K|}{3} \|c\|_{0,\infty,K} \left|(\nabla u_h^L)|_K\right| . \tag{5.27}$$

If $u_h^L < 0$ on K, then (5.27) holds true, since the left-hand side is negative. Thus, repeating arguments from (5.26), we obtain

$$(cu_h^L, \varphi_i)_K \le C_2 \sum_{E \in \mathcal{E}(P_i)} h_E^3 \|c\|_{0,\infty,\omega(E)} \left|[\![\nabla u_h^L]\!]_E\right| . \tag{5.28}$$

Now, we consider the bubble part of u_h. To this end, we test (5.14) with $\varphi_K^B \in V_{h0}$ and get for $f = 0$

$$\begin{aligned} &D(\nabla u_h^B, \nabla \varphi_K^B)_K + (\boldsymbol{b} \cdot \nabla u_h^B, \varphi_K^B)_K + (cu_h^B, \varphi_K^B)_K + S_h(u_h^B, \varphi_K^B) \\ &= -D(\nabla u_h^L, \nabla \varphi_K^B)_K - (\boldsymbol{b} \cdot \nabla u_h^L, \varphi_K^B)_K - (cu_h^L, \varphi_K^B)_K - S_h(u_h^L, \varphi_K^B) \\ &= -(\boldsymbol{b} \cdot \nabla u_h^L, \varphi_K^B)_K - (cu_h^L, \varphi_K^B)_K \end{aligned}$$

since (5.22) and $\kappa_h(\nabla u_h^L) = 0$ are satisfied. Consequently, we obtain for $u_h^B|_K = u_K \varphi_K^B$ by integrating by parts

$$u_K = -\frac{(\boldsymbol{b} \cdot \nabla u_h^L, \varphi_K^B) + (cu_h^L, \varphi_K^B)}{(D + \tau_K)\, |\varphi_K^B|_{1,K}^2 + \left(c - \frac{1}{2}\nabla \cdot \boldsymbol{b}, (\varphi_K^B)^2\right)_K} \tag{5.29}$$

which together with (5.26), (5.28) and $|\varphi_K^B|_{1,K} \sim 1$ yields

$$|u_K| \le C_3(b, c) \frac{h_K^2}{D + \tau_K} \left|(\nabla u_h^L)|_K\right| . \tag{5.30}$$

Taking $\tau_K \sim h_K$ into consideration, we end up with

$$\begin{aligned} (\boldsymbol{b} \cdot \nabla u_h^B, \varphi_i) &= \sum_{K \subset \Omega_i} u_K (\boldsymbol{b} \cdot \nabla \varphi_K^B, \varphi_i)_K \le \sum_{E \in \mathcal{E}(P_i)} C_4(\boldsymbol{b}, c) h_E^2 \left|[\![\nabla u_h^L]\!]\right| , \\ (cu_h^B, \varphi_i) &= \sum_{K \subset \Omega_i} u_K (c \varphi_K^B, \varphi_i)_K \le \sum_{E \in \mathcal{E}(P_i)} C_5(\boldsymbol{b}, c) h_E^3 \left|[\![\nabla u_h^L]\!]\right| . \end{aligned} \tag{5.31}$$

Finally, we have for the stabilising term

$$S_h(u_h, \varphi_i) = 0 \tag{5.32}$$

since $\kappa_h(\nabla \varphi_i) = 0$. From (5.26), (5.28), (5.31) and (5.32) we conclude the assertion (5.25). □

Now, we are able to state the main result concerning the discrete minimum principle property (DMPP).

Theorem 5.8 *Let $f = 0$. Then for sufficiently large $c_p > 1$ the semilinear form*

$$A(u_h; v_h) = a(u_h, v_h) + S_h(u_h, v_h) + j_h(u_h; v_h)$$

satisfies discrete minimum principle property from Definition (5.2).

Proof. Let u_h attain a negative local minimum at an inner node P_i. Then we have

$$j_h(u_h; \varphi_i) = -c_p \sum_{E \in \mathcal{E}(P_i)} \gamma_{1,E}(\boldsymbol{b}, c) h_E^2 \big| [\![\nabla u_h^L]\!]_E \big| . \tag{5.33}$$

Then, it follows from Lemma 5.5 and 5.7 the estimate

$$A(u_h; \varphi_i) \leq - \sum_{E \in \mathcal{E}(P_i)} \left\{ \frac{D}{2} h_E + (c_p - 1)\gamma_{1,E}(\boldsymbol{b}, c) h_E^2 \right\} \big| [\![\nabla u_h^L]\!]_E \big| . \tag{5.34}$$

Setting

$$\alpha_E := \frac{D}{2} h_E + (c_p - 1)\gamma_{1,E}(\boldsymbol{b}, c) h_E^2 ,$$

we get $\alpha_E > 0$ for sufficiently large $c_p > 1$. Thus, $A(\cdot;\cdot)$ satisfies the discrete minimum principle property from Definition 5.2. □

5.3 Linear convergence of edge oriented shock capturing scheme

Lemma 5.9 *Let u_h be the solution of* (5.14) *and let $u \in H^2(\Omega)$ be solution of* (5.4). *Then, it holds for $\tau_k \sim h_K$, $D \sim 1$ and sufficiently small h*

$$|||u - u_h||| \leq C h . \tag{5.35}$$

Proof. Let $\tilde{u}_h$ be solution of linear scheme (5.8). Starting with V_h-coercivity of the bilinear form $a(\cdot,\cdot) + S_h(\cdot,\cdot)$, we can follow

$$\begin{aligned} |||\tilde{u}_h - u_h||| &\leq \sup_{v_h \in V_{h0}} \frac{a(\tilde{u}_h - u_h, v_h) + S_h(\tilde{u}_h - u_h, v_h)}{|||v_h|||} = \sup_{v_h \in V_{h0}} \frac{j_h(u_h; v_h)}{|||v_h|||} \\ &= \sup_{v_h \in V_{h0}^L} \frac{j_h(u_h; v_h)}{|||v_h|||} \end{aligned} \tag{5.36}$$

due to the fact that $\frac{\partial v_h}{\partial t_E} = 0$ holds for all $v_h \in V_{h0} \setminus V_{h0}^L$. Next, we apply Cauchy–Schwarz inequality in order to bound $j_h(u_h; v_h)$

$$j(u_h, v_h) \le C \left(\sum_{E \in \mathcal{E}_h} h_E^3 \left| [\![\nabla u_h^L]\!] \right|^2 \right)^{1/2} \left(\sum_{E \in \mathcal{E}_h} h_E^3 |v_h|^2_{1,\infty,\omega(E)} \right)^{1/2} .$$

Employing scaling arguments and using the fact $(\nabla u_h^L, \nabla u_h^B) = 0$, we get consequently

$$j(u_h, v_h) \le Ch|u_h|_1 |v_h|_1 .$$

Combining this with (5.36), we conclude from $|||v_h||| \ge D^{1/2}|v_h|_1$ and (5.10) the estimate

$$\begin{aligned} |||\tilde{u}_h - u_h||| &\le C\frac{h}{\sqrt{D}}|u_h|_1 \le C\frac{h}{D}|||u_h - \tilde{u}_h||| + C\frac{h}{D}|||u - \tilde{u}_h||| + C\frac{h}{D}|||u||| \\ &\le C\frac{h}{\sqrt{D}}|||u_h - \tilde{u}_h||| + C\frac{h^2}{D}|u|_2 + C\frac{h}{D}|||u||| \end{aligned}$$

from which we infer for $D \sim 1$ and sufficiently small h

$$|||\tilde{u}_h - u_h||| \le Ch. \tag{5.37}$$

Using the triangle inequality

$$|||u - u_h||| \le |||\tilde{u}_h - u_h||| + |||u - \tilde{u}_h|||$$

and a priori estimate (5.10) together with (5.37), we end up the proof. □

Our numerical tests show that the solution u_h of the scheme (5.14) converges to the solution u of (5.4) linearly also in the convection dominating case $D \ll |\boldsymbol{b}|$ and for $D \ll h$.

5.4 Numerical tests

In this section we present some numerical result for the shock capturing method for local projection stabilisation applied to convection-diffusion-reaction problem.

We perform computations on the unit square $\Omega = (0,1)^2$. The coarser triangular mesh is refined regularly. The discretised nonlinear problems are linearised by fixed point iteration. Let $u_h^0 \in V_h$ be some prescribed initial solution. In each iteration step $k \in \mathbb{N}$ we solve the following discrete problem

Find $u_h^{(k+1)} \in V_h$ with $u_h^{(k+1)}|_\Gamma = g_h$ such that

$$a(u_h^{(k+1)}, v_h) + S_h(u_h^{(k+1)}, v_h) = (f, v_h) - \tilde{j}_h(u_h^{(k)}, v_h) \quad \forall\, v_h \in V_{h0} \tag{5.38}$$

The operator $\tilde{j}_h$ stands for the smoothed nonlinear shock capturing operator j_h which is created by the following approximation of the discontinuous function

$$\operatorname{sign}(t) \approx \tanh(t/\mu)$$

where $\mu > 0$ is a suitable chosen regularisation parameter. The above procedure results in the sequence of the linear systems

$$\boldsymbol{A}_h \boldsymbol{x}_h^{(k+1)} = \boldsymbol{F}_h(\boldsymbol{x}_h^{(k)}), \quad k = 0, 1, 2, \dots$$

where the stiffness matrix $\boldsymbol{A}_h$ corresponds to the Galerkin and LPS parts, the right hand side $\boldsymbol{F}_h$ corresponds to the right hand side of (5.38) and is updated in each iteration step. The solution vector in the iteration step k is denoted by $\boldsymbol{x}_h^{(k+1)}$. In our computations we set $\mu = 1.0$ and abort the nonlinear iteration process if the euclidian norm satisfies $\|\boldsymbol{x}_h^{(k+1)} - \boldsymbol{x}_h^{(k)}\| \leq 1e-6$. For small values of perturbation parameter $D > 0$ we use damping procedure in order to enforce the convergence of the iteration (5.38).

5.4.1 Numerical study of convergence for smooth solution

At first, we investigate the rate of convergence for the problem with

$$\boldsymbol{b} = (0.1, 0)^T, \quad c = 1$$

in the diffusion dominating ($D = 1.0$) and convection dominating ($D = 1e-7$) cases. The right hand side f is such that the exact solution is given by

$$u(x,y) = \exp\left(-\frac{(x-0.5)^2}{a_w} - \frac{3(y-0.5)^2}{a_w}\right),$$

with parameter $a_w = 0.2$, see Figure 5.1.

The coarsest mesh of Friedrichs-Keller type consists of 8 triangular cells and the number of degrees of freedom (dof) on each level is given in Table 5.1. In Tables 5.2 and 5.3

Table 5.1: Total number of degrees of freedom.

level	0	1	2	3	4	5	6	7
dofs	17	57	209	801	3137	12417	49409	197121

we report errors and rates of convergence for $D = 1.0$. For both methods the orders of convergence are almost the same.

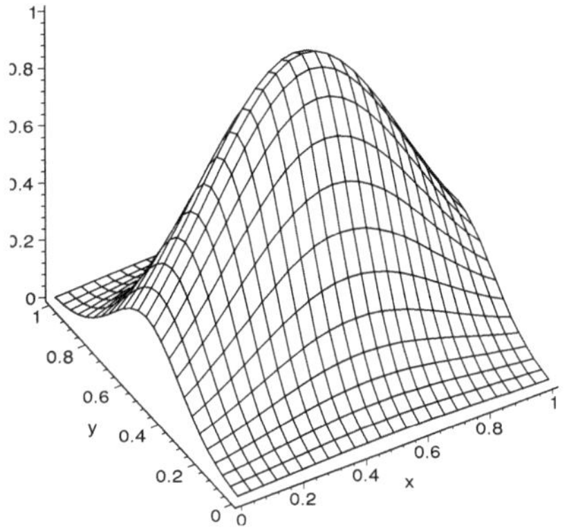

Figure 5.1: The Gaussian as exact solution.

Table 5.2: Errors and rates of convergence for LPS solution, $D = 1.0$.

level	error and rate					
	$\|\|u - \tilde{u}_h\|\|_0$		$\|u - \tilde{u}_h\|_1$		$\|\|\|u - \tilde{u}_h\|\|\|$	
0	1.082e-1		9.902e-1		9.988e-1	
1	6.092e-2	0.829	7.526e-1	0.396	7.558e-1	0.402
2	1.857e-2	1.714	4.310e-1	0.804	4.316e-1	0.808
3	4.819e-3	1.947	2.196e-1	0.973	2.197e-1	0.974
4	1.216e-3	1.986	1.103e-1	0.993	1.103e-1	0.994
5	3.048e-4	1.997	5.522e-2	0.998	5.523e-2	0.998
6	7.624e-5	1.999	2.762e-2	1.000	2.762e-2	1.000
7	1.906e-5	2.000	1.381e-2	1.000	1.381e-2	1.000

Table 5.3: Errors and rates of convergence for LPS solution with shock capturing, $D = 1.0$.

level	error and rate					
	$\|\|u - u_h\|\|_0$		$\|u - u_h\|_1$		$\|\|\|u - u_h\|\|\|$	
0	1.083e-1		9.902e-1		9.988e-1	
1	6.092e-2	0.829	7.527e-1	0.396	7.559e-1	0.402
2	1.858e-2	1.714	4.310e-1	0.804	4.316e-1	0.808
3	4.819e-3	1.947	2.196e-1	0.973	2.197e-1	0.974
4	1.216e-3	1.986	1.103e-1	0.993	1.103e-1	0.994
5	3.048e-4	1.997	5.522e-2	0.998	5.523e-2	0.998
6	7.624e-5	1.999	2.762e-2	1.000	2.762e-2	1.000
7	1.906e-5	2.000	1.381e-2	1.000	1.381e-2	1.000

Table 5.4: Errors and rates of convergence for LPS solution, $D = 1e-7$.

level	error and rate					
	$\|\|u-\tilde{u}_h\|\|_0$		$\|u-\tilde{u}_h\|_1$		$\|\|\|u-\tilde{u}_h\|\|\|$	
0	8.501e-2		1.116e+0		1.166e-1	
1	3.909e-2	1.121	9.306e-1	0.262	5.470e-2	1.092
2	8.833e-3	2.146	5.011e-1	0.893	1.770e-2	1.628
3	2.023e-3	2.126	2.461e-1	1.026	5.932e-3	1.577
4	4.936e-4	2.035	1.227e-1	1.005	2.053e-3	1.531
5	1.226e-4	2.010	6.134e-2	1.000	7.187e-4	1.514
6	3.059e-5	2.003	3.069e-2	0.999	2.529e-4	1.507
7	7.641e-6	2.001	1.535e-2	1.000	8.923e-5	1.503

Table 5.5: Errors and rates of convergence for LPS solution with shock capturing, $D = 1e-7$.

level	error and rate					
	$\|\|u-u_h\|\|_0$		$\|u-u_h\|_1$		$\|\|\|u-u_h\|\|\|$	
0	1.981e-1		1.307e+0		2.150e-1	
1	1.473e-1	0.427	1.267e+0	0.044	1.519e-1	0.5012
2	6.834e-2	1.108	7.545e-1	0.748	6.980e-2	1.122
3	3.490e-2	0.969	3.913e-1	0.947	3.528e-2	0.984
4	1.750e-2	0.996	1.987e-1	0.978	1.760e-2	1.004
5	8.754e-3	0.999	1.061e-1	0.905	8.780e-3	1.003
6	4.380e-3	0.999	6.058e-2	0.809	4.387e-3	1.001
7	2.174e-3	1.011	3.678e-2	0.720	2.176e-3	1.012

The results from Tables 5.4 and 5.5 indicate that the order of convergence for the errors in L^2 and LPS norm are reduced in the convection dominating case ($D = 1e-7$) by one and by one half compared to the LPS method, respectively.

In the following we plot usually only the linear part of solutions, otherwise we specify the plotted values.

5.4.2 Skew flow problem with exponential and internal layers

We start our numerical study for problems possessing boundary layers of different kinds. First, we apply LPSSC method to the skew flow problem with the following constant data

$$\boldsymbol{b} = (-\sin\alpha, -\cos\alpha)^T, \quad c = 0, \quad f = 0$$

and discontinuous boundary condition

$$g(x,y) = \begin{cases} 0 & \text{for} \quad 0 \le x \le 1,\ y = 0 \\ 0 & \text{for} \quad x = 1,\ 0 \le y < 0.75 \\ 1 & \text{otherwise}\,. \end{cases}$$

The discontinuity at the boundary is transported in the direction of the convection field $\boldsymbol{b}$ until reaching the boundary part with zero condition since the diffusion parameter is very small. The solution possesses a steep gradient and thus exhibits a boundary layer in the region of discontinuity and at the part of the outlet boundary

$$\{(x,y): \quad 0 \le x \le 1\,, \quad y = 0\} \subset \Gamma\,.$$

We observe that the solution stabilised by the local projection method is stable away from the boundary layers. However, it exhibits still spurious oscillations in the region of the exponential boundary layer at the part of the outflow boundary which has been mentioned above, see Figures 5.2-5.15.

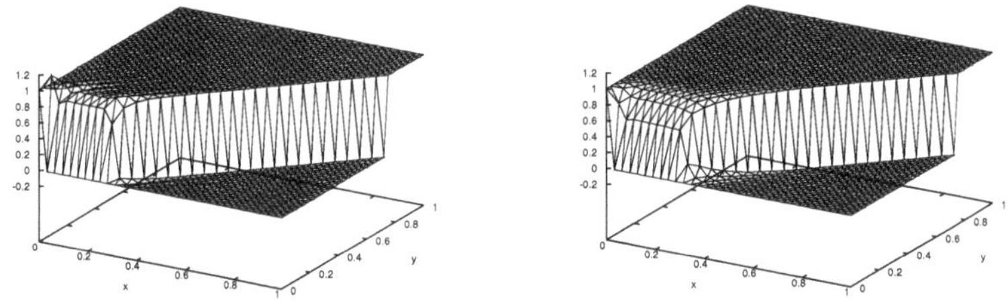

Figure 5.2: Solutions of skew flow problem obtained by LPS (left) and (right), $\tau_K = 0.01 h_K$, $\alpha = 45\,^\circ$.

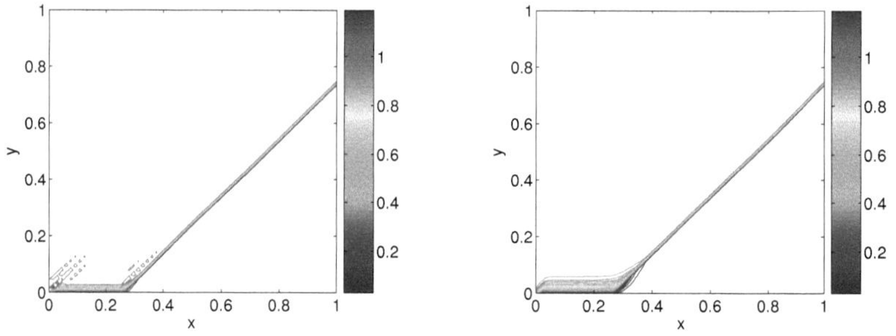

Figure 5.3: Contour lines of solution obtained by LPS (left) and LPSSC (right), $\tau_K = 0.01h_K$, $\alpha = 45^\circ$.

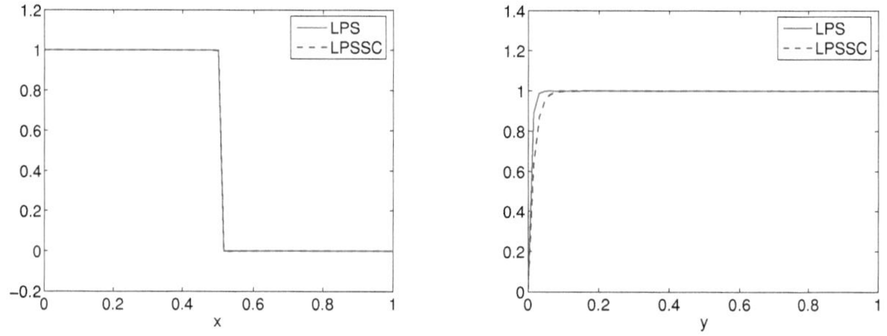

Figure 5.4: Profiles of solution at $y = 0.25$ (left) and $x = 0.125$ (right), $\alpha = 45^\circ$, $\tau_K = 0.01h_K$.

In Figure 5.8 we plot sections of the whole solutions $\tilde{u}_h$ and u_h obtained by LPS and LPSSC methods, respectively. The smoothing effect of the jump operator j_h can be well observed

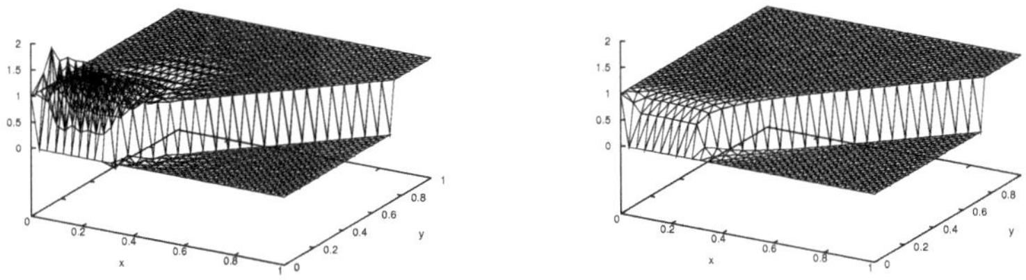

Figure 5.5: Solutions of skew flow problem obtained by LPS (left) and (right), $\tau_K = 0.1h_K$, $\alpha = 45^\circ$.

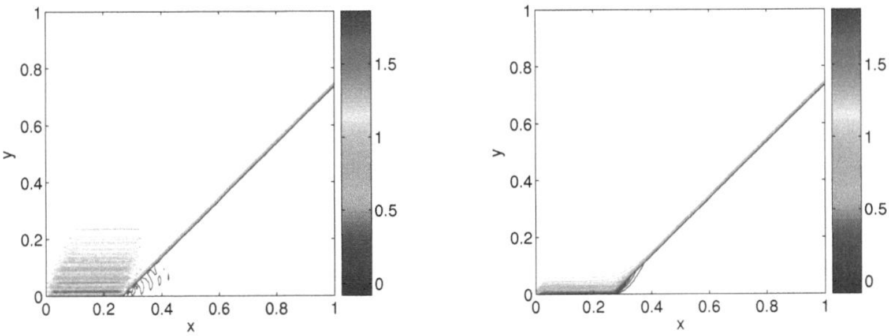

Figure 5.6: Contour lines of solution obtained by LPS (left) and LPSSC (right), $\tau_K = 0.1h_K$, $\alpha = 45^\circ$.

for $\alpha = 45^\circ$ and $\alpha = 55^\circ$. The choice of the stabilisation parameter $\tau_k = 0.01h_K$ seems to be quite good if no additional edge stabilisation is applied, see Figures 5.2 and 5.4 for $\alpha = 45^\circ$ and Figures 5.9 and 5.11 for $\alpha = 55^\circ$. The discrete maximum principle is much more violated if $\tau_K = 0.1h_K$, see Figures 5.5 and 5.7 for $\alpha = 45^\circ$ and Figures 5.12 and 5.14 for $\alpha = 55^\circ$.

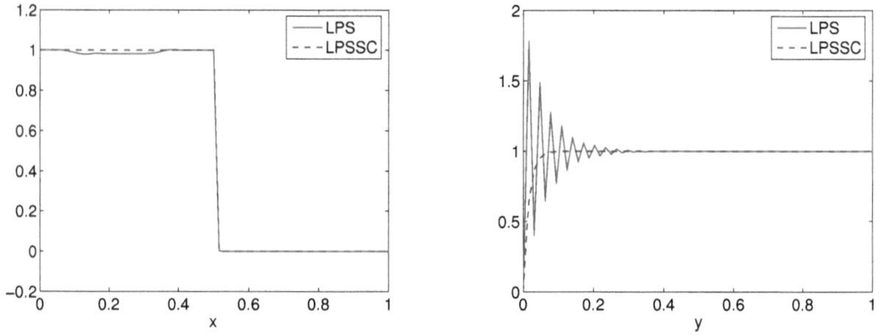

Figure 5.7: Profiles of solution at $y = 0.25$ (left) and $x = 0.125$ (right), $\alpha = 45\,^\circ$, $\tau_K = 0.1 h_K$.

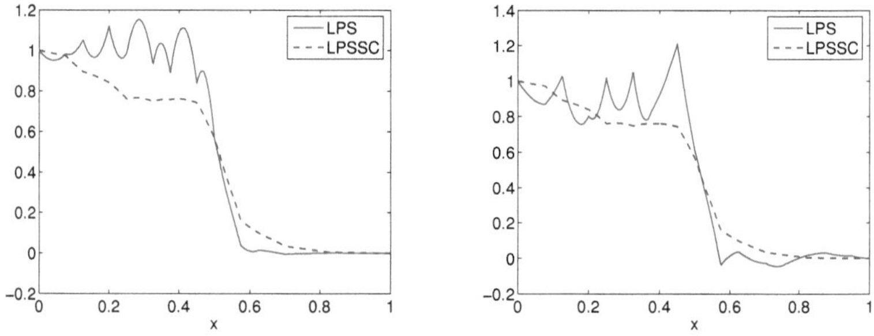

Figure 5.8: Full profiles of $\tilde{u}_h(x, \frac{13}{24})$ and $u_h(x, \frac{13}{24})$ on level 2, $\alpha = 45\,^\circ$, $\tau_K = 0.01 h_K$ (left), $\tau_K = 0.1 h_K$ (right).

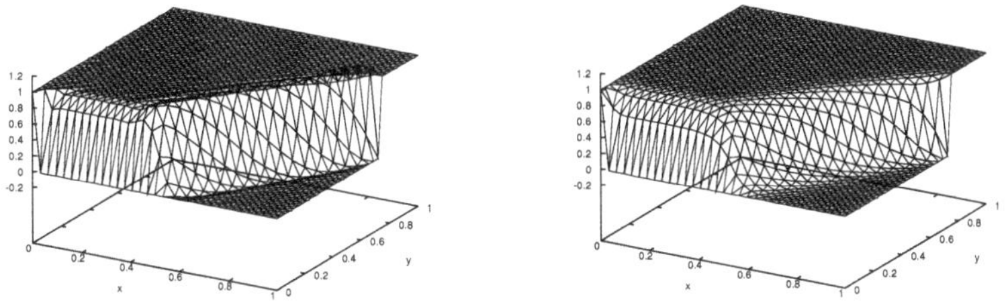

Figure 5.9: Solutions of skew flow problem obtained by LPS (left) and (right), $\tau_K = 0.01h_K$, $\alpha = 55^\circ$.

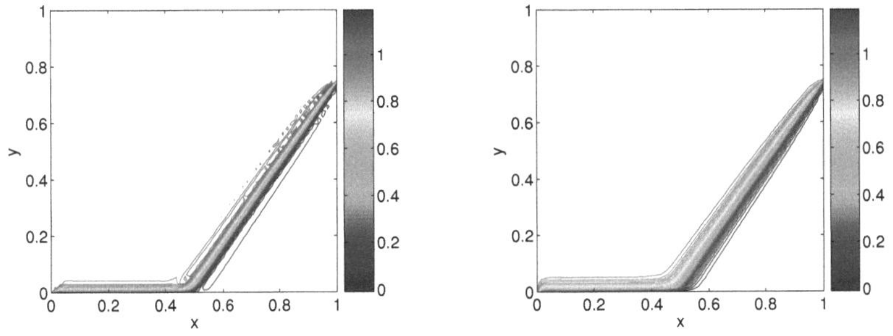

Figure 5.10: Contour lines of solutions obtained by LPS (left) and LPSSC (right), $\tau_K = 0.01h_K$, $\alpha = 55^\circ$.

5.4.3 Rotating flow problem with exponential and interior layers

Our next problem is a benchmark for problems with an interior layer and an exponential layer. It has boundary conditions of mixed type. Let

$$D = 10^{-7}, \quad \boldsymbol{b} = \left(8xy(1-x), -4(2x-1)(1-y^2)\right)^T, \qquad c = 0$$

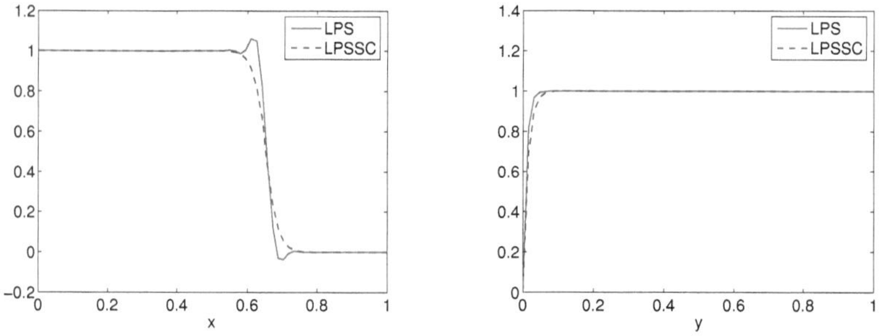

Figure 5.11: Profiles of solution at $y = 0.25$ (left) and $x = 0.125$ (right), $\alpha = 55^\circ$, $\tau_K = 0.01h_K$.

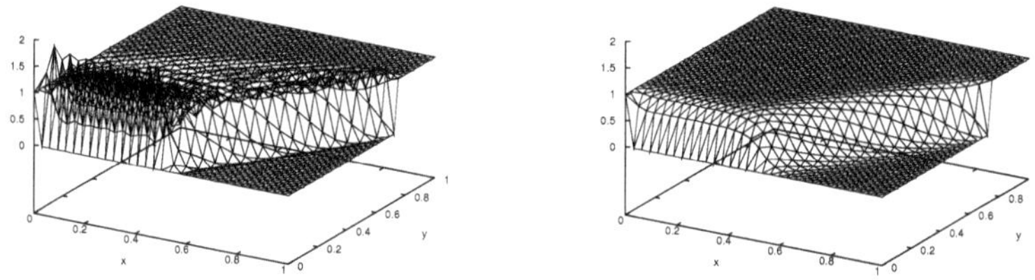

Figure 5.12: Solutions of skew flow problem obtained by LPS (left) and (right), $\tau_K = 0.1h_K$, $\alpha = 55^\circ$.

and

$$\Gamma_N := \{(x,y) \in \partial\Omega \,:\, 1/2 < x < 1,\, y = 0\}, \qquad \Gamma_D := \partial\Omega \setminus \Gamma_N.$$

We prescribe on the Dirichlet boundary Γ_D the piecewise constant function

$$g_D(x,y) = \begin{cases} 1 & \text{for } 1/4 \le x \le 1/2\,,\ y = 0, \\ 1 & \text{for } 0 \le y \le 1\,,\ x = 1, \\ 0 & \text{otherwise,} \end{cases}$$

while the homogeneous Neumann condition $g_N = 0$ will be used on Γ_N. The right hand side in (5.1) is given by $f = 0$. The streamlines of the convection field $\boldsymbol{b}$ are shown in Figure 5.16. We observe that the unstabilised Galerkin solution exhibits non-physical oscillations over the whole domain Ω and therefore is completely useless, see Figure 5.17. Applying a first order upwind stabilisation, we obtain a stable solution. However, the

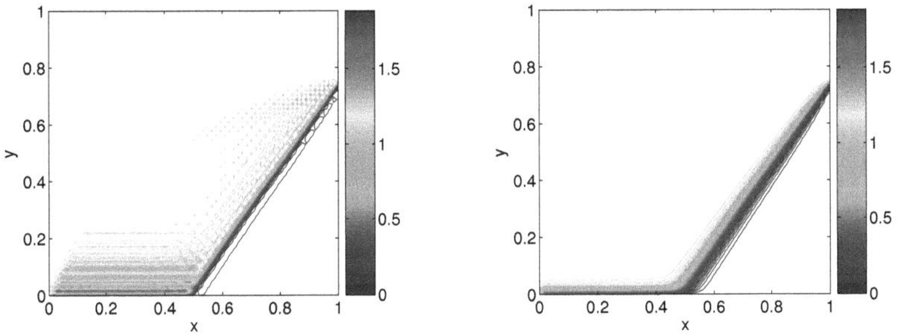

Figure 5.13: Contour lines of solutions obtained by LPS (left) and LPSSC (right), $\tau_K = 0.1h_K$, $\alpha = 55^\circ$

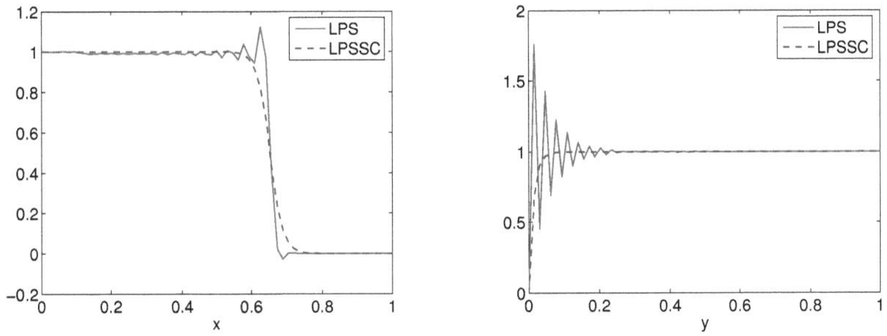

Figure 5.14: Profiles of solution at $y = 0.25$ (left) and $x = 0.125$ (right), $\alpha = 55^\circ$, $\tau_K = 0.1h_K$.

sharp inflow profile at $\{(x,y): \quad x \geq 1/4, \ y = 0\}$ is smeared out, see Figure 5.21. The discrete solution obtained by LPSSC method satisfies the discrete maximum principle, see Figures 5.18 and 5.20.

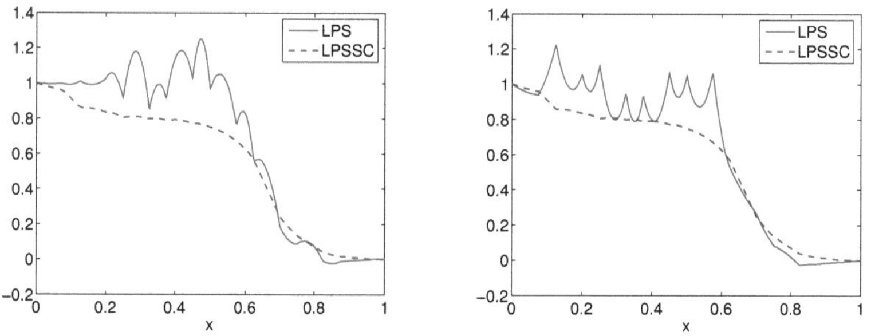

Figure 5.15: Full profiles of $\tilde{u}_h(x, \frac{13}{24})$ and $u_h(x, \frac{13}{24})$ on the coarse level 2, $\alpha = 55^\circ$, $\tau_K = 0.01h_K$ (left), $\tau_K = 0.1h_K$ (right).

5.4.4 Solution with parabolic layer

The solution of our last example exhibits two parabolic boundary layers. Let

$$D = 10^{-7}, \quad \boldsymbol{b} = (0, 1 + x^2)^T, \quad c = 0$$

and

$$\Gamma_N := \{(x, y) \in \partial\Omega : 0 < x < 1, \, y = 1\}, \qquad \Gamma_D := \partial\Omega \setminus \Omega_N.$$

We use homogeneous Neumann condition $g_N = 0$ on Γ_N while the Dirichlet boundary condition g_D on Γ_D is given by

$$g_D = \begin{cases} 1 & \text{for } 0 \le x \le 1, \; y = 0, \\ 1 - y & \text{otherwise.} \end{cases}$$

Furthermore, the right hand side of (5.1) is $f = 0$. The solution of (5.1) exhibits parabolic layers at the vertical walls $x = 0$ and $x = 1$. The discrete solution obtained by LPS violates the discrete maximum principle whereas the discrete solution obtained by LPSSC satisfies it well, see Figures 5.22 and 5.24.

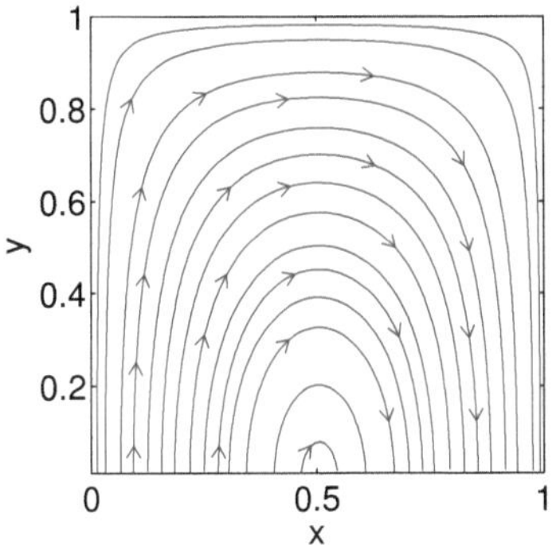

Figure 5.16: Streamlines of the convection field $\boldsymbol{b}$

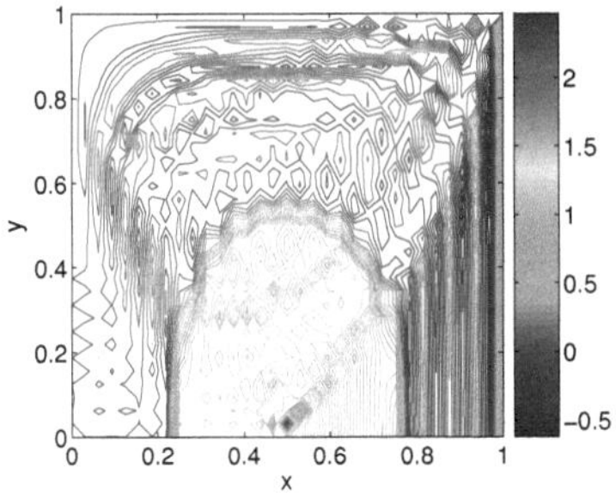

Figure 5.17: Galerkin solution obtained without any stabilisation

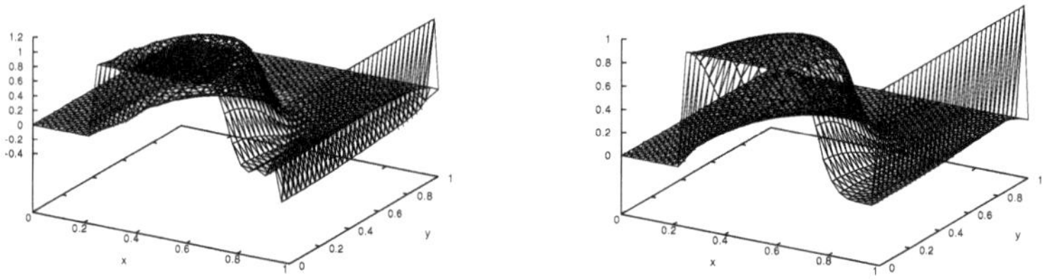

Figure 5.18: Solution of rotating flow problem, LPS (left) and LPSSC (right), $\tau_K = 0.01h_K$.

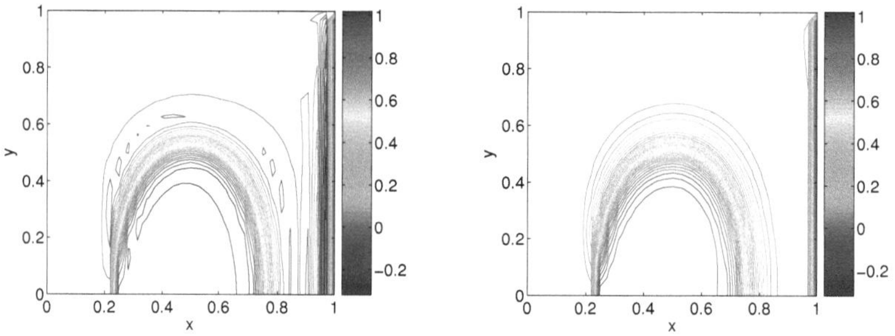

Figure 5.19: Contour lines of solution obtained by LPS (left) and LPSSC (right), $\tau_K = 0.01h_K$.

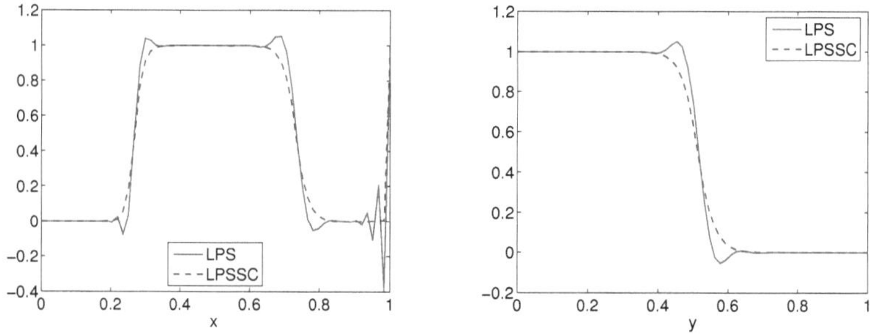

Figure 5.20: Profiles of solution, at $x = 0.5$ (left) and at $y = 0.25$ (right) obtained by LPS and LPSSC, $\tau_K = 0.01h_K$.

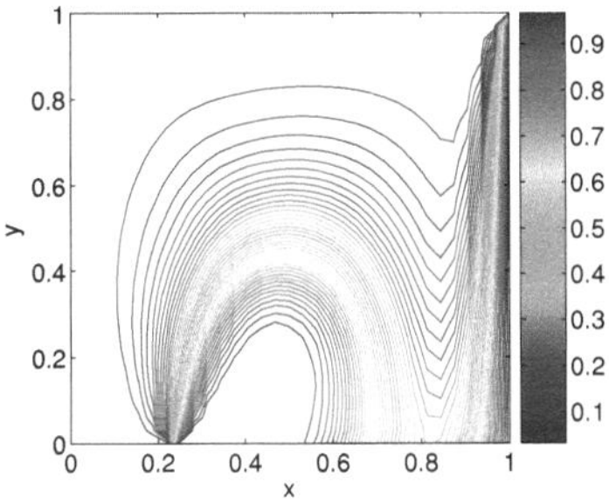

Figure 5.21: Solution obtained by upwind stabilisation with the smeared out the inflow front at $\{(x,y): \quad x \geq 1/4,\ y = 0\}$.

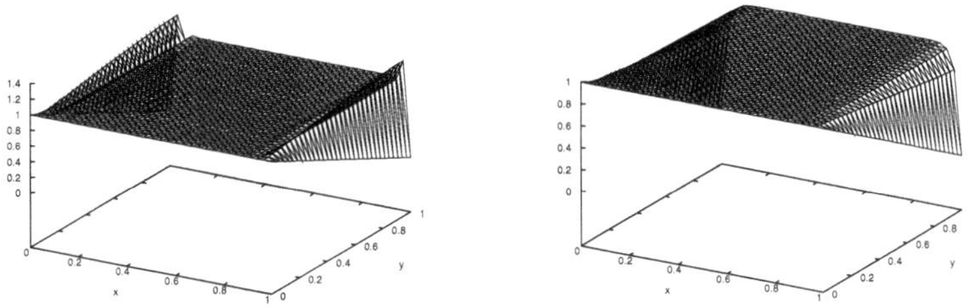

Figure 5.22: Solution of parabolic flow problem, LPS (left) and LPSSC (right), $\tau_K = 0.01h_K$.

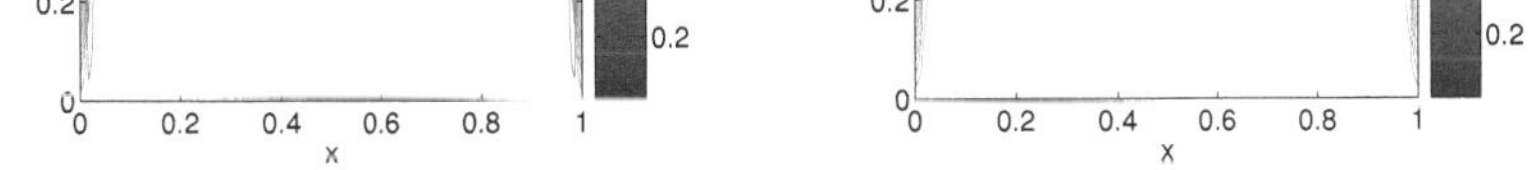

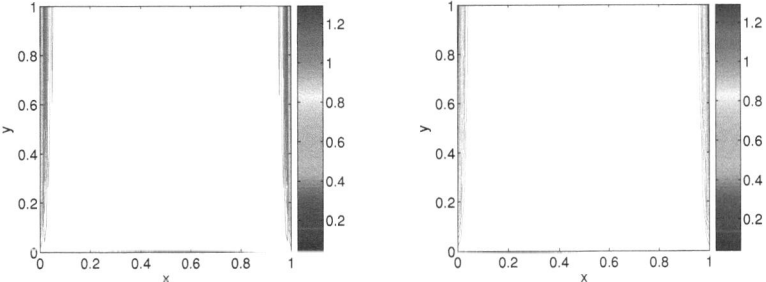

Figure 5.23: Contour lines of solution obtained by LPS (left) and LPSSC (right), $\tau_K = 0.01h_K$.

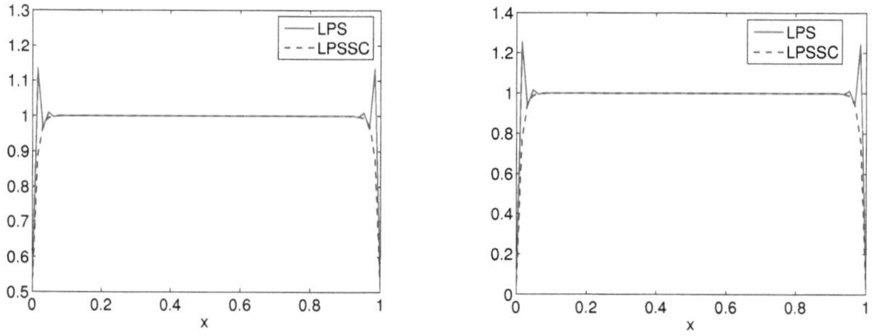

Figure 5.24: Profiles of solution, at $y = 0.5$ (left) and at $y = 1$ (right) obtained by LPS and LPSSC, $\tau_K = 0.01h_K$.

6 Summary

The question how to establish stable and accurate discretisation methods for singularly perturbed problems belongs to the challenging problems of numerical mathematics. This work deals with a stable finite element discretisation of Dirichlet problem for extended Brinkman–Forchheimer equations in two and three dimensional domains. The governing equations describe flow dynamics in fixed bed reactors. This type of reactors is widely used in chemical engineering. Since the considered model problem is rare, it has not been intensively studied until now. Its nonlinear character and nonstandard form of differential operators occurring in the equations require a special handling. There are three "daemons" which we have forced to get back:

- nonlinearity
- inf-sup instability (improper choice of approximation spaces)
- high Reynolds numbers (convection dominance)

We present analytical results concerning the existence of solution in suitable Sobolev spaces using abstract theory of saddle point problems. To this end, we adapt Hopf extension known from the theory of Navier–Stokes equations and apply a version of Schauder's fixed point theorem.

After the analytical part, we establish stability and a priori estimates for finite element solutions. The key point in the proof of stability is the discrete inf-sup condition for the bilinear form which is nonstandard in our case. This is the second daemon which occurs when the approximation spaces for the velocity and pressure are improperly chosen. Employing the patch techniques of Boland–Nicolaides, we show that the choice of a certain family of finite element pairs with discontinuous pressure approximation leads on quadrilaterals/hexahedrons to stable and accurate solutions in the case of low Reynolds numbers.

Another "daemon" in the fluid dynamics is the case of high Reynolds numbers. We present a stabilisation method for equal order elements which are not inf-sup stable. Using the one-level variant of the local projection stabilisation, we are able to obtain stable solutions for the velocity and pressure. Based on the published results from the unified convergence analysis for Oseen equations, we prove again the optimal order of convergence for the linearised model equations.

After controlling three "daemons", we present a method for enhancing the accuracy of the finite element solution. Using the fact that the error of the special finite interpolant to the finite element solution is of one order better on the axis parallel grids, we can apply sophisticated extrapolation techniques in order to increase the accuracy of the computed finite element solution. The already published superconvergence results for three dimensional Navier–Stokes equations can be successfully applied to our model. We show that the superclose estimates for the nonstandard bilinear forms hold in two dimensions and we estimate errors caused by the nonlinear terms. Applying post-processing to the inf-sup stable pair of the continuous piecewise biquadratic velocity and discontinuous piecewise linear pressure, we get a numerical solution which is convergent of one order better.

Plenty of numerical tests and simulations of flows in fixed bed and packed bed membrane reactors are performed in order to verify our theoretical results. The proposed schemes produce stable and accurate solutions and can be successfully applied on desktop computers.

In the last part of our work we enhance the stability of the low order local projection scheme for scalar convection-diffusion-reaction problems. While the solution obtained by the local projection method is stable away of boundary layer, it can still exhibit spurious oscillations at the boundary layer. This can be called as "the last revenge" of the third daemon. To avoid the undesired instability we propose an edge oriented shock capturing scheme for triangular meshes. Here we let the first and the third daemon clash with each other. Finally, the nonlinear "daemon" hidden in the shock capturing term can be banished using a simple fixed point iteration. We prove that the finite element solution obtained from our scheme satisfies a discrete maximum principle and is at least linearly convergent in the diffusion-dominated case. Our numerical results are in a good agreement with the developed theory.

Bibliography

[1] R. A. Adams. *Sobolev Spaces.* Pure and applied mathematics. Academic Press, 1995.

[2] L. El Alaoui, A. Ern, and E. Burman. A nonconforming finite element method with face penalty for advection-diffusion equations. In *Numerical mathematics and advanced applications*, pages 512–519. Springer, Berlin, 2006.

[3] T. Apel. *Anisotropic finite elements. Local estimates and applications.* Advances in Numerical Mathematics. Teubner, Leipzig, 1999.

[4] D. N. Arnold, D. Boffi, and R. S. Falk. Approximation by quadrilateral finite elements. *Math. Comput.*, 71(239):909–922, 2002.

[5] R. Becker and M. Braack. A finite element pressure gradient stabilization for the Stokes equations based on local projections. *Calcolo*, 38(4):173–199, 2001.

[6] R. Becker and M. Braack. A two-level stabilization scheme for the Navier–Stokes equations. In M. Feistauer et al., editor, *Numerical mathematics and advanced applications*, pages 123–130, Berlin, 2004. Springer-Verlag.

[7] C. Bernardi and B. Raugel. Analysis of some finite elements of the Stokes problem. *Math. Comp.*, 44:71–79, 1985.

[8] Christine Bernardi, Frédéric Laval, Brigitte Métivet, and Bernadette Pernaud-Thomas. Finite element approximation of viscous flows with varying density. *SIAM J. Numer. Anal.*, 29(5):1203–1243, 1992.

[9] O. Bey. *Strömungsverteilung und Wärmetransport in Schüttungen.* Nummer 570 in Fortschritt-Berichte, VDI Reihe 3, Verfahrenstechnik. Düsseldorf: VDI Verlag, 1998.

[10] J. Boland and R. Nicolaides. Stability of finite elements under divergence constrains. *SIAM J. Numer. Anal.*, 20(4):722–731, 1983.

[11] M. Braack and E. Burman. Local projection stabilization for the Oseen problem and its interpretation as a variational multiscale method. *SIAM J. Numer. Anal.*, 43(6):2544–2566, 2006.

[12] M. Braack, E. Burman, V. John, and G. Lube. Stabilized finite element methods for the generalized Oseen problem. *Comput. Methods Appl. Mech. Eng.*, 196(4–6):853–866, 2007.

[13] J. Brandts and M. Křížek. History and future of superconvergence in three-dimensional finite element methods. In P. Neittaanmäki et al., editor, *Finite element methods. Three-dimensional problems*, volume 15 of *Math. Sci. and Appl., GAKUTO International Series*, pages 22–33, Tokyo, Gakkotosho, 2001.

[14] Jan H. Brandts, Sergey Korotov, and Michal Křížek. The discrete maximum principle for linear simplicial finite element approximations of a reaction-diffusion problem. *Linear Algebra Appl.*, 429(10):2344–2357, 2008.

[15] E. Burman. A unified analysis for conforming and nonconforming stabilized finite element methods using interior penalty. *SIAM J. Numer. Anal.*, 43(5):2012–2033 (electronic), 2005.

[16] Erik Burman and Alexandre Ern. Nonlinear diffusion and discrete maximum principle for stabilized Galerkin approximations of the convection–diffusion-reaction equation. *Comput. Methods Appl. Mech. Engrg.*, 191(35):3833–3855, 2002.

[17] Erik Burman and Alexandre Ern. Stabilized Galerkin approximation of convection-diffusion-reaction equations: discrete maximum principle and convergence. *Math. Comp.*, 74(252):1637–1652 (electronic), 2005.

[18] P. G. Ciarlet and P.-A. Raviart. Maximum principle and uniform convergence for the finite element method. *Comput. Methods Appl. Mech. Engrg.*, 2:17–31, 1973.

[19] P. Clément. Approximation by finite element functions using local regularization. *RAIRO Anal. Numer.*, 9:77–84, 1975.

[20] Ramon Codina. A discontinuity-capturing crosswind-dissipation for the finite element solution of the convection-diffusion equation. *Comput. Methods Appl. Mech. Engrg.*, 110(3-4):325–342, 1993.

[21] Andrei Drăgănescu, Todd F. Dupont, and L. Ridgway Scott. Failure of the discrete maximum principle for an elliptic finite element problem. *Math. Comp.*, 74(249):1–23 (electronic), 2005.

[22] S. Ergun. Fluid flow through packed columns. *Chemical Engineering Progress*, 48(2):89–94, 1952.

[23] A. Ern and J.-L. Guermond. *Theory and practice of finite elements*, volume 159 of *Applied Mathematical Sciences*. Springer-Verlag, New York, 2004.

[24] Alexandre Ern and Jean-Luc Guermond. *Theory and practice of finite elements*, volume 159 of *Applied Mathematical Sciences*. Springer-Verlag, New York, 2004.

[25] Lawrence C. Evans. *Partial differential equations*, volume 19 of *Graduate Studies in Mathematics*. American Mathematical Society, Providence, RI, 1998.

[26] George J. Fix, Max D. Gunzburger, and Janet S. Peterson. On finite element approximations of problems having inhomogeneous essential boundary conditions. *Comput.*

Math. Appl., 9(5):687–700, 1983.

[27] L. P. Franca. An overview of the residual-free-bubbles method. In *Numerical methods in mechanics (Concepción, 1995)*, volume 371 of *Pitman Res. Notes Math. Ser.*, pages 83–92. Longman, Harlow, 1997.

[28] L. P. Franca and A. Russo. Deriving upwinding, mass lumping and selective reduced integration by residual-free bubbles. *Appl. Math. Lett.*, 9(5):83–88, 1996.

[29] L. P. Franca and L. Tobiska. Stability of the residual free bubble method for bilinear finite elements on rectangular grids. *IMA J. Numer. Anal.*, 22(1):73–87, 2002.

[30] V. Girault and P.-A. Raviart. *Finite Element Methods for Navier–Stokes Equations. Theorie and Algorithms.* Springer-Verlag, 1986.

[31] P. Grisvard. *Elliptic problems in nonsmooth domains*, volume 24 of *Monographs and Studies in Mathematics.* Pitman (Advanced Publishing Program), Boston, MA, 1985.

[32] J.-L. Guermond. Stabilization of Galerkin approximations of transport equations by subgrid modeling. *M2AN*, 33(6):1293–1316, 1999.

[33] Max D. Gunzburger and Janet S. Peterson. On conforming finite element methods for the inhomogeneous stationary Navier-Stokes equations. *Numer. Math.*, 42(2):173–194, 1983.

[34] Antti Hannukainen, Sergey Korotov, and Tomáš Vejchodský. Discrete maximum principle for FE solutions of the diffusion-reaction problem on prismatic meshes. *J. Comput. Appl. Math.*, 226(2):275–287, 2009.

[35] V. Heuveline and F. Schieweck. H^1-interpolation on quadrilateral and hexahedral meshes with hanging nodes. *Computing*, 80(3):203–220, 2007.

[36] V. Heuveline and F. Schieweck. On the inf-sup condtion for higher order mixed fem on meshes with hanging nodes. *M2AN*, 41(1):1–20, 2007.

[37] E. Hopf. Ein allgemeiner Endlichkeitssatz der Hydrodynamik. *Math. Ann.*, 117:764–775, 1941.

[38] U. Hornung. *Homogenization and Porous Media.* Springer-Verlag, 1997.

[39] T. J. R. Hughes. Multiscale phenomena: Green's functions, the Dirichlet–to–Neumann formulation, subgrid scale models, bubbles and the origins of stabilized methods. *Comput. Methods Appl. Mech. Eng.*, 127(1-4):387–401, 1995.

[40] T. J. R. Hughes and A. Brooks. A multidimensional upwind scheme with no crosswind diffusion. In *Finite element methods for convection dominated flows (Papers, Winter Ann. Meeting Amer. Soc. Mech. Engrs., New York, 1979)*, volume 34 of *AMD*, pages 19–35. Amer. Soc. Mech. Engrs. (ASME), New York, 1979.

[41] T. J. R. Hughes and G. Sangalli. Variational multiscale analysis. projection, optimization, the fine-scale Greens' function, and stabilized methods. USNCCM8, Austin July, 27-29, 2005.

[42] T. J. R. Hughes and G. Sangalli. Variational multiscale analysis: The fine-scale Green's function, projection, optimization, localization, and stabilized methods. Technical Report 05-46, Institute for Computational Engineering and Sciences, University of Texas at Austin, 2005.

[43] Tsutomu Ikeda. *Maximum principle in finite element models for convection-diffusion phenomena*, volume 4 of *Lecture Notes in Numerical and Applied Analysis*. Kinokuniya Book Store Co. Ltd., Tokyo, 1983. North-Holland Mathematics Studies, 76.

[44] Volker John and Petr Knobloch. On spurious oscillations at layers diminishing (SOLD) methods for convection-diffusion equations. I. A review. *Comput. Methods Appl. Mech. Engrg.*, 196(17-20):2197–2215, 2007.

[45] Volker John and Petr Knobloch. On spurious oscillations at layers diminishing (SOLD) methods for convection-diffusion equations. II. Analysis for P_1 and Q_1 finite elements. *Comput. Methods Appl. Mech. Engrg.*, 197(21-24):1997–2014, 2008.

[46] Volker John and Gunar Matthies. MooNMD—a program package based on mapped finite element methods. *Comput. Vis. Sci.*, 6(2-3):163–169, 2004.

[47] P. N. Kaloni and Jianlin Guo. Steady nonlinear double-diffusive convection in a porous medium based upon the Brinkman-Forchheimer model. *J. Math. Anal. Appl.*, 204(1):138–155, 1996.

[48] János Karátson and Sergey Korotov. An algebraic discrete maximum principle in Hilbert space with applications to nonlinear cooperative elliptic systems. *SIAM J. Numer. Anal.*, 47(4):2518–2549, 2009.

[49] János Karátson, Sergey Korotov, and Michal Křížek. On discrete maximum principles for nonlinear elliptic problems. *Math. Comput. Simulation*, 76(1-3):99–108, 2007.

[50] Petr Knobloch. Numerical solution of convection-diffusion equations using upwinding techniques satisfying the discrete maximum principle. In *Proceedings of Czech-Japanese Seminar in Applied Mathematics 2005*, volume 3 of *COE Lect. Note*, pages 69–76. Kyushu Univ. The 21 Century COE Program, Fukuoka, 2006.

[51] Petr Knobloch and Lutz Tobiska. On the stability of finite-element discretizations of convection-diffusion-reaction equations. *IMA Journal of Numerical Analysis*, pages 1–18 (electronic), 2009. DOI:10.1093/imanum/drp020.

[52] D. Kuzmin, M. J. Shashkov, and D. Svyatskiy. A constrained finite element method satisfying the discrete maximum principle for anisotropic diffusion problems. *J. Comput. Phys.*, 228(9):3448–3463, 2009.

[53] Q. Lin. Superclose FE-theory becomes a table of integrals. In *Finite Element Methods—superconvergence, post-processing and a posteriori estimates*, pages 217–225. Marcel Dekker Inc., New York, 1998.

[54] Q. Lin, J. Li, and A. Zhou. A rectangle test for the Stokes problem. In *Proceedings of Systems Science & Systems Engineering*, pages 240–241. Great Wall(H.K.) Culture Publish Co., 1991.

[55] Q. Lin and J. Pan. Global superconvergence for rectangular elements in the Stokes problem. In *Proceedings of Systems Science & Systems Engineering*, pages 371–378. Great Wall(H.K.) Culture Publish Co., 1991.

[56] Q. Lin and N. Yan. High efficient finite elements (in Chinese). Publ. of Hebei University, 1996.

[57] Q. Lin, N. Yan, and A. Zhou. A rectangle test for interpolated finite elements. In *Proc. of Sys. Sci. & Sys Engrg. Great Wall(H.K)*, pages 217–229. Culture Publ. Co., 1991.

[58] Qun Lin, Lutz Tobiska, and Aihui Zhou. Superconvergence and extrapolation of non-conforming low order finite elements applied to the Poisson equation. *IMA J. Numer. Anal.*, 25(1):160–181, 2005.

[59] J. L. Lions. *Quelques methodes de resolution des problemes aux limites non lineaires.* Dunod/Gauthier-Villars, Paris, 1969.

[60] Gunar Matthies. Mapped finite elements on hexahedra. necessary and sufficient conditions for optimal interpolation errors. *Numer. Algorithms*, 27(4):317–327, 2001.

[61] Gunar Matthies, Piotr Skrzypacz, and Lutz Tobiska. Superconvergence of a 3D finite element method for stationary Stokes and Navier–Stokes problems. *Numer. Methods Partial Differential Equations*, 21(4):701–725, 2005.

[62] Gunar Matthies, Piotr Skrzypacz, and Lutz Tobiska. A unified convergence analysis for local projection stabilisations applied to the Oseen problem. *M2AN Math. Model. Numer. Anal.*, 41(4):713–742, 2007.

[63] Gunar Matthies, Piotr Skrzypacz, and Lutz Tobiska. Stabilization of local projection type applied to convection-diffusion problems with mixed boundary conditions. *Electron. Trans. Numer. Anal.*, 32:90–105, 2008.

[64] Gunar Matthies and Lutz Tobiska. The inf-sup condition for the mapped Q_k-P_{k-1}^{disc} element in arbitrary space dimensions. *Computing*, 69(2):119–139, 2002.

[65] Pingbing Ming and Zhong-Ci Shi. Quadrilateral mesh. *Chinese Ann. Math. Ser. B*, 23(2):235–252, 2002. Dedicated to the memory of Jacques-Louis Lions.

[66] Pingbing Ming and Zhong-Ci Shi. Quadrilateral mesh revisited. *Comput. Methods Appl. Mech. Engrg.*, 191(49-50):5671–5682, 2002.

[67] L.A. Oganesjan and L.A. Ruchovec. An investigation of the rate of convergence of variation-difference schemes. *Z. Vycisl. Mat. i Mat. Fiz.*, 9:1102–1120, 1969.

[68] Alfio Quarteroni and Alberto Valli. *Numerical approximation of partial differential equations*. Springer-Verlag, Berlin, 1994.

[69] Hans-Görg Roos, Martin Stynes, and Lutz Tobiska. *Robust numerical methods for singularly perturbed differential equations*, volume 24 of *Springer Series in Computational Mathematics*. Springer-Verlag, Berlin, second edition, 2008. Convection-diffusion-reaction and flow problems.

[70] L. R. Scott and S. Zhang. Finite element interpolation of nonsmooth functions satisfying boundary conditions. *Math. Comput.*, 54(190):483–493, 1990.

[71] Andreas Seidel-Morgenstern (ed.). *Membrane Reactors*. Wiley-VCH, Berlin, 1 edition, 2010. Distributing reactants to Improve Selectivity and Yield.

[72] Piotr Skrzypacz. Superkonvergenz von finite Elemente Methoden für skalare elliptische Gleichungen und für die stationären Stokes- und Navier–Stokes–Probleme. Diplomarbeit, Institut für Analysis und Numerik, Otto-von-Guericke-Universität Magdeburg, Germany, 2002.

[73] Pavel Šolín and Tomáš Vejchodský. A weak discrete maximum principle for *hp*-FEM. *J. Comput. Appl. Math.*, 209(1):54–65, 2007.

[74] Martin Stynes and Lutz Tobiska. Using rectangular Q_p elements in the SDFEM for a convection-diffusion problem with a boundary layer. *Appl. Numer. Math.*, 58(12):1789–1802, 2008.

[75] L. Tobiska. Analysis of a new stabilized higher order finite element method for advection–diffusion equations. *Comput. Methods Appl. Mech. Eng.*, 196(1–3):538–550, 2006.

[76] Evangelos Tsotsas. *Über die Wärme- und Stoffübertragung in durchströmten Festbetten: Experimente, Modelle, Theorien*. VDI-Verl., Düsseldorf, 1990.

[77] T. Vejchodský and P. Šolín. Discrete maximum principle for a 1D problem with piecewise-constant coefficients solved by *hp*-FEM. *J. Numer. Math.*, 15(3):233–243, 2007.

[78] Tomáš Vejchodský and Pavel Šolín. Discrete maximum principle for higher-order finite elements in 1D. *Math. Comp.*, 76(260):1833–1846 (electronic), 2007.

[79] Tomáš Vejchodský and Pavel Šolín. Discrete maximum principle for Poisson equation with mixed boundary conditions solved by *hp*-FEM. *Adv. Appl. Math. Mech.*, 1(2):201–214, 2009.

[80] L.B. Wahlbin. *Superconvergence in Galerkin Finite Element Methods*. Number 1605 in Lecture Notes in Math. Springer-Verlag, 1995.

[81] Jinchao Xu and Ludmil Zikatanov. A monotone finite element scheme for convection-diffusion equations. *Math. Comp.*, 68(228):1429–1446, 1999.

Index

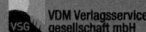

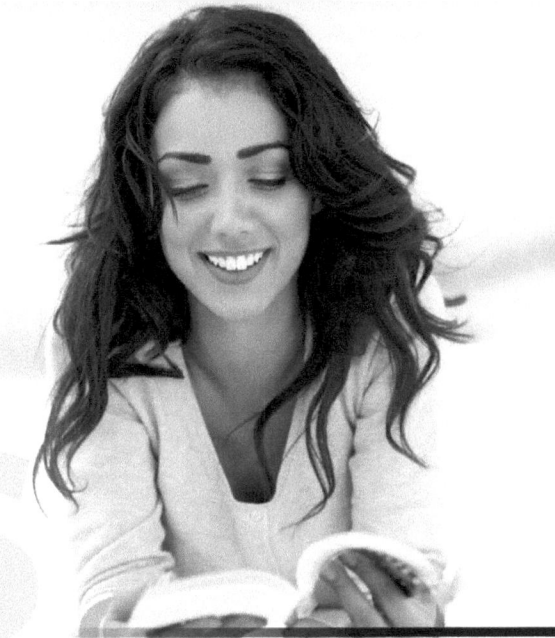

MIX
Papier aus verantwortungsvollen Quellen
Paper from responsible sources
FSC® C105338

Printed by Books on Demand GmbH, Norderstedt / Germany